Synthese Library

Studies in Epistemology, Logic, Methodology, and Philosophy of Science

Volume 482

The aim of *Synthese Library* is to provide a forum for the best current work in the methodology and philosophy of science and in epistemology, all broadly understood. A wide variety of different approaches have traditionally been represented in the Library, and every effort is made to maintain this variety, not for its own sake, but because we believe that there are many fruitful and illuminating approaches to the philosophy of science and related disciplines.

Special attention is paid to methodological studies which illustrate the interplay of empirical and philosophical viewpoints and to contributions to the formal (logical, set-theoretical, mathematical, information-theoretical, decision-theoretical, etc.) methodology of empirical sciences. Likewise, the applications of logical methods to epistemology as well as philosophically and methodologically relevant studies in logic are strongly encouraged. The emphasis on logic will be tempered by interest in the psychological, historical, and sociological aspects of science. In addition to monographs *Synthese Library* publishes thematically unified anthologies and edited volumes with a well-defined topical focus inside the aim and scope of the book series. The contributions in the volumes are expected to be focused and structurally organized in accordance with the central theme(s), and should be tied together by an extensive editorial introduction or set of introductions if the volume is divided into parts. An extensive bibliography and index are mandatory.

Lorenzo Azzano

Dispositional Reality

A Novel Approach to Power Ontology and Metaphysics

Springer

Lorenzo Azzano
University of Santiago de Compostela
Santiago de Compostela, Spain

ISSN 0166-6991 ISSN 2542-8292 (electronic)
Synthese Library
ISBN 978-3-031-52624-4 ISBN 978-3-031-52625-1 (eBook)
https://doi.org/10.1007/978-3-031-52625-1

This Springer imprint is published by the registered company Springer Nature Switzerland AG
The registered company address is: Gewerbestrasse 11, 6330 Cham, Switzerland

Paper in this product is recyclable.

Acknowledgments

The idea behind this book was first conceived during the Spring 2020 lockdown. At the time, I was trying to develop the version of explanatory dispositionalism introduced in my previous papers, and I was fascinated by the possibility of slimming down my metaphysical baggage. I was thinking about metaphysical structure and perspicuity a lot during that period, even if in the beginning I did not know how to precisely conceptualize them for the goals I had set for myself. There were a couple of false starts, before reaching the framework that would ultimately appear in this manuscript.

Much has happened since then, and I have been fortunate enough to benefit from the comments and discussions of many colleagues and friends. Firstly, I received useful feedback on early stages of the project by Andrea Borghini, Giacomo Giannini, and Simone Gozzano. Early in 2021, I was beneficiary of a *Juan de la Cierva* grant at the University of Santiago de Compostela, working within the *Episteme* research group under the supervision of José Luis Falguera López. The manuscript would have been in much worse shape – if any at all –, without my stay in Santiago. The research group's interest in ontological and meta-ontological deflationism made a last impression on my position. Therefore, I want to thank all the good people in Santiago: besides José Luis, Concha Martínez-Vidal, Xavier De Donato Rodríguez, Martin Pereira-Fariña, Ismael Ordóñez Miguéns, Violeta Conde Borrego, and Alejandro Gracia De Rienzo. Early versions of Chapters 2 and 3 were used as a basis for a graduate course in Santiago; I would like to thank everyone who attended and made it for a lively discussion.

Other portions of the book were presented in conferences and workshops throughout these years: besides Santiago (in more than one occasion), Salamanca, Tilburg, Berlin, and Bristol. I would like to thank everyone who attended these talks and provided some feedback. Additionally, some people read longer portions of various drafts to provide extensive comments: Giorgio Lando, Barbara Vetter, and Lisa Vogt. My thanks go to them as well; and I also must thank a particularly incisive reviewer whose comments have immensely helped the manuscript. My gratitude also goes to Andrea Raimondi, an inexhaustible source of discussion and support throughout these complicated years. As they say, I alone am responsible for everything written in the following pages. Last but not least, I also would like to thank Otávio Bueno, the most patient and helpful editor I could have hoped for.

This book is part of a trajectory that began my years ago. I remember being an undergraduate student, discussing the existence of dispositions with my supervisor Andrea Bottani, when he asked me a question that ultimately boiled down to "what if I am a nominalist?". I didn't exactly know how to answer then, but I do now. In a way, this book is my reply. I am heavily indebted to Andrea, who was the first to ignite my interest in analytic ontology and metaphysics. I am also indebted to Massimiliano Carrara for the many suggestions and discussions throughout the years.

Research for this manuscript has counted on funding from FEDER (*Ministerio de Ciencia, Innovación y Universidades – Agencia Estatal de Investigación*) by means of grant PID2020-115482GB-I00 ("Deflationary views in ontology and metaontology"), and also the grant ED431B2023/24 received by the *Episteme* research group from the *Xunta de Galicia*.

Finally, this book is dedicated to Valeria – though my name appears on the cover, we are not separate existences.

Abstract

Dispositionalism, a popular variant of non-Humean metaphysics, submits that dispositions, powers, or capacities are part of the furniture of the world. In this book, I advance an original approach to dispositionalism revolving around the notion of Dispositional Reality; the novelty lies in the fact that the account, unlike most alternatives on the market, does not require the reification of objects, facts, properties, nor their dispositional essences, and is in fact compatible with a more deflationary approach to dispositions, while still being true to the non-Humean spirit of the proposal. This power metaphysics without powers allows one to dispel several puzzles in recent literature, or recast them under a new light. Albeit with its own peculiarities, this proposal constitutes a variant of explanatory dispositionalism, according to which realism about dispositions ought not to be understood as an ontological inflation, but as an explanatory inversion within the nomic and modal family. Some of these explanations are hereby attempted, and a study of various types of non-causal explanation is provided.

Contents

Chapter 1
Introduction

Some dispositional talk is not merely true, but metaphysically perspicuous, thus revealing the presence of a dispositional aspect of reality – what I call Dispositional Reality. The remainder of this book is devoted to spelling out this claim, and offering (partial) motivation for it. Before embarking on this task, however, there's something we ought to make clear: my underlying motivation, and the issues I find more pressing on the matter – my angle, so to speak. This is what this introductory chapter is about.

The opening sentence may come off as quite cryptic. Here is a quick Q&A, in no specific order. First question: what do you mean with "dispositional aspect of reality"? Do you mean something along the lines of "dispositional properties", as friends of dispositions usually do? Answer: no. Dispositional Reality is not just a fancy stand-in for dispositional properties. As a matter of fact, Dispositional Reality is not a thing at all, nor any collection of things. Second question: what is it, then? Answer: you'll see. Third question: what is it for a sentence to be not merely true, but metaphysically perspicuous? Answer: you'll see. Fourth (and final) question: what do you mean with "dispositional talk", anyway? Answer: nothing too specific, really. I simply mean those claims which ascribe dispositions, powers, or capacities to entities in the physical world. Dispositional talk, so understood, is no philosopher's artifice; on the contrary, it abounds in all facets of life, from everyday contexts to scientific ones. Some people are irascible, in the sense that they are predisposed to get angry in certain situations. Sulphur matches' heads are flammable, in the sense that they are disposed to easily ignite at ambient temperature, whereas sugar is water-soluble, in the sense that it is disposed to dissolve when immersed in water. Hydrogen peroxide has the power to oxidize other substances. Liquid neon has an extraordinary cooling capacity. An electrically charge particle is disposed to undergo a force $\vec{F} = q\vec{E}$ in an electric field $\vec{E}$. So on and so forth. There's no point harping on issues of grammar or logical form for the time being: all of these ascriptions characterize objects according to certain potentials they possess, according to what they can do, or would do in certain circumstances; although, of course, such cursory

L. Azzano, *Dispositional Reality*, Synthese Library 482,
https://doi.org/10.1007/978-3-031-52625-1_1

remarks run the risk of being circular, or "only [a] clumsy and inexact linguistic gestures to dispositions" (Martin, 2007, p. 19), they can still function to circumscribe a *prima facie* subject matter.[1]

What should be the philosopher's attitude, vis-à-vis this overabundance of dispositions? Roughly put, one can either be a friend or a foe of dispositions: in the broadest possible formulation, a friend of dispositions (a dispositionalist) claims that dispositions are part of the furniture of the world: thus, these dispositional ascriptions tell us something important – for lack of a better expression – about the make-up of reality. Foes of dispositions reject that.

This extremely coarse-grained distinction can be refined in a variety of different ways, and dispositionalism is now a limited yet stable presence in contemporary metaphysics and philosophy of science; whereas a complete genealogy and taxonomy of the debate is neither possible nor desirable, we will explore part of this literature in this book. In the summary provided in Azzano (2019), for example, I submitted that the question of realism for dispositions is usually framed as a debate around the existence of certain properties, or their essence, or their fundamentality, or their truth-making aptness vis-à-vis certain classes of sentences. These loosely contiguous characterizations, as different as they are, are usually backed by shared assumptions about the kind of endeavour dispositionalism really is. First and foremost, the assumption usually is that dispositions constitute an ontology; they are properties, or something in the vicinity thereof. Not only that: they are *dispositional* properties. Perhaps the thought is that this privileged subclass of properties is essentially dispositional, whose nature separate them from other kinds of properties; and perhaps this inherent dispositional nature is what makes them apt to function as truth-makers in the way the dispositionalist desires; or perhaps the thought is that dispositional properties are fundamental properties, in the sense that they can be found in the smallest and most fundamental interstices of nature, as friends of dispositions sometimes like to argue.

At any rate, this is where power metaphysics, as a recent subdiscipline of metaphysics, flourishes with its own peculiar brand of puzzles and epicycles. For once these dispositional properties are admitted into the ontology, questions about them multiply: are dispositional properties related to their manifestations? In what way? In other words, what is the nature of the apparently mysterious directedness of dispositions? And what kind of properties are we dealing with, exactly? Are dispositional properties tropes? Are they universals? And if so, of what kind? What makes a property dispositional, anyway? Are all properties dispositional? Are properties exhaustively dispositional, in some sense of the word? And if properties are dispositional in the sense that they have a dispositional essence, what can we say about this essence? Is this supposedly dispositional essence tied to counterfactuals? How? Is it structural? Qualitative? Primitive? Relatedly, are

[1] What about abilities? This will be the job for another time. What about what's beyond the physical world, if there are such things? This will also be the job for another time.

dispositional properties simple? Or do they encompass both a qualitative and a dispositional aspect? How can that be?

This web of strictly interconnected questions is but a portion of the countless issues that friends of dispositions have tackled in recent years (this is not meant as a disparaging remark: I include myself amongst them). But I will argue: in our enthusiasm for dispositionalism, perhaps we bought too much. We were lured, or perhaps wrong-footed, into disputes that didn't need discussing, at least not at this stage. To borrow an expression from Mondadori and Morton (1976), we ate a "poisoned pawn". This curious expression was originally meant to signal the kind of fallacious reasoning according to which realism and objectivity in a certain subject matter can only be achieved by paying an ontological price, viz. by committing oneself to the existence of, and then perhaps discussing the nature of, entities of a certain kind. In Chap. 2, when discussing the meta-ontological backdrop of my proposal, I will dub a very similar kind of reasoning an "ontological fallacy". Mondadori and Morton introduced the concept to discuss the complicated relation between realism about modality and modal realism. But the same concept, or at least a very similar one, ought in my opinion be discussion in relation to the question of realism for dispositions. The admission fee for the dispositionalists' club ought not be paid with the coin of ontology. For there is, on the face of it, a far less committing kind of realism concerning dispositionality than the one usually assumed: it's the idea that (some) dispositional ascriptions are metaphysically revealing, in the sense that they are telling us something important about the structure and make-up of reality; viz. they are revealing us dispositional aspect of reality. Whether that takes the form of properties, facts, or what have you, is immaterial at this stage: a nominalist about properties might very well believe the most perspicuous description of reality to be dispositional, and thus for reality to be "dispositional-ish".

The reader may object: this is all too vague. This is a fair point. As a matter of fact, I agree: these are vague claims. But they needn't be. The contemporary metaphysician now has at their disposal many useful tools that can help them synthetize this nebulous gesture into an (hopefully) coherent metaphysics. This is the goal we are currently setting out to achieve: in this book, I will deploy and develop many such metaphysical tools in a somewhat integrated framework, to articulate an account of Dispositional Reality as a minimal nucleus for dispositionalism. If there are further reasons to delve into some of the many issues listed above, to complement or expand the account perhaps, so be it; but it is not required. Alternatively, it is also possible, as I will also do, to pursue a kind of dispositionalism purposefully designed as more ontologically and metaphysically deflationary than the vast majority of its competitors. This is not because I believe there to be inherent virtues to deflationism (whether that is true, I suspect, cannot be decided at such a degree of generality), nor for the mere sake of antagonism: this is only because – in opposition with the far more inflationary proposal presently on the market – a more encompassing logical space of positions can be more easily delineated.

To be sure, the account of Dispositional Reality emerging from these pages is indeed in opposition to most other proposals in the field, a veritable power metaphysics without powers which does not require (and may very well exclude) the reification of dispositions, properties, facts, or even objects. Within the context of my account of Dispositional Reality, many of the most cryptic conundrums concerning dispositions simply deflate and disappear, whereas others can be observed under a new light, and resolved in a completely new way.

This approach is also meant to recapture, in an ontologically deflated way, the distinction between "metaphysically innocent" as opposed to "loaded" dispositional talk, a distinction that has recently been put under the spotlight by dispositionalists. In a sense, everyone agrees that there are dispositions; indeed, a once-influential paper by Michael Fara, for all intents and purposes a foe of dispositions, begins with the very sentence "[o]bjects have dispositions" (Fara, 2005, p. 43). But in another sense, it is only dispositionalists who believe in dispositions; this distinction is often marked by a lexicographical distinction, albeit, unfortunately, not always the same one: the courageous reader will face a veritable flurry of terms-of-art in the following pages. A popular lexicographical distinction (albeit by no means the only one, nor the orthodox one) is that everyone believes that objects have "dispositions", whereas only dispositionalists believe in "powers"; although one usually gestures at a non-descript relation of metaphysical grounding or relative fundamentality between them – so as to distinguish derivative (and lightweight) dispositions from fundamental (and heavyweight) powers – it is still unclear what the exact difference is there between dispositions and powers, nor what relation is there between the two; this is where the danger of ontological inflation becomes particularly worrying.

On my account, this distinction is simply deflated as a difference in attitude concerning dispositional talk: we can take dispositional talk to be true, and then we can take it to be perspicuous. There's no need to stratify the realm of dispositions in any way. This approach has more general and systematic virtues: the rejection of error-theory, a standard move by foes of dispositions, is now embedded in a more systematic meta-metaphysical framework.

The second crucial element of my account of Dispositional Reality is the *explanatory* element. This is meant to recapture some of the initial impetus for power metaphysics, not as an ontological inflation (there wasn't, and still isn't, a shared understanding of what these "powers" are that we are admitting into the ontology, no metaphysical framework to characterize them, let alone a formal theory), but as a kind of explanatory inversion concerning items in the nomic and modal family, the kind of inversion that was perhaps first glimpsed in Goodman's (1954, p. 43) observation that "the problem of dispositions is really simpler than the problem of counterfactuals". Contemporary friends of dispositions usually think of a world inherently imbued with nomic and modal features, which needn't be super-imposed by external laws of nature, nor outsourced through possible worlds. Yet, again, we might be overstepping our boundaries if we simply assumed that a world enriched with nomic and modal features ought to be a world inhabited by a *sui generis* kind of nomic or modal entities – in this case, powers-as-entities. Again, a poisoned pawn. On the contrary, the claim might simply be that nomic or modal talk is successfully explained through dispositional talk. The kind of explanation I have in mind here is primarily (but not exclusively, as we will see), what is occasionally called constitutive explanation; successful constitutive explanations involve an increase in perspicuity from the *explanandum* to the *explanans*; e.g. if we constitutively explained instances of *grue* through time-indexed instances of *green* and *blue*, that could be accompanied by the claim that grue-talk is less perspicuous than green-talk and blue-talk. Similarly, by providing explanations for a class of nomic and modal sentences through a class of dispositional sentences, we would also be claiming that nomic and modal talk is less perspicuous than dispositional talk, in the sense that, say, nomic talk is a less perspicuous way to describe the same portions of reality that dispositional talk is describing more perspicuously. Talk of laws, of counterfactuals, of what is possible or necessary, is but an imperfect gesture towards Dispositional Reality (here is again where I am tiptoeing around vague and potentially compromising language; this book will hopefully vindicate this approach). Some of these explanations will be attempted in this book. Others will have to wait.

Thus, my account of Dispositional Reality qualifies as a kind of explanatory dispositionalism. Explanatory dispositionalism is not an entirely novel phenomenon, as *per* Azzano (2019, 2020), and Vetter (2020). Similarly, grounding theories of powers have very recently been submitted which bear some resemblance to what is being claimed here (Coates, 2020, 2022; Kimpton-Nye, 2021; Tugby, 2021, 2022). But only indirectly: as the reader will see, I have much to object to standard characterizations of metaphysical grounding. Furthermore, by virtue of its more deflationary elements, my account may also be loosely related to the nominalism-friendly strand of dispositionalism from Shoemaker (1980), Whittle (2009), and Vogt (2022); once again, the differences amongst these accounts, and between these accounts and mine, abound. I will deal with this literature in due time.

Incidentally, this is an excellent opportunity to introduce some of the issues of philosophical methodology guiding the rest of the book. Traditionally, many discussions concerning the reality of dispositions revolved around counterexamples to the so-called conditional analysis of dispositions; troubles concerning the conditional analysis, in this context, mostly revolved around interference counterexamples violating the material adequacy of these bridge-principles. A vast literature now exists on these topics, but a certain line of thought has emerged amongst dispositionalists that, even if all the counterexamples could be resolved, that wouldn't *per se* provide an argument against dispositionalism; as a matter of fact, the availability of these principles is only indirectly related to the question of realism for dispositions. With some qualification, I agree with this sentiment. Once matter of realisms are established (as I suggest) as matters of metaphysical perspicuity, one can see how the ready-made availability of materially adequate bridge principles between two classes of sentences cannot, on its own, establish the reality of anything: for nothing in this bridge principle itself dictates a direction of increased perspicuity. Bridging grue-talk with green/blue talk is easy enough, but it does not tell us whether grue is real (or natural, or fundamental, or what have you), as opposed to green or blue. The problem is not alleviated by calling this bridge-principle an "analysis", or "reduction", or "explanation" (as in the case of dispositions and the so-called conditional analysis): without a reason for choosing a direction as opposed to another, this is but an empty gesture. This is also true for our supposed "explanations" of laws and modality in terms of disposition. That there are supposedly explanatory bridge-principles linking dispositional talk to nomic or modal talk tells us nothing about the direction of metaphysical enquiry (perhaps, as in the case of Bird (2007a, p. 46), the explanation comes in the form of a derivation; but if the troubles of the DN model of explanation thought us anything, is that deductions too go both ways).

Alternatively, one could try to read an explanatory direction in such bridge-principles by observing whichever kind of determinative relation is underpinning the explanation itself. But assuming that non-causal explanation and determinations are at play here (dispositions do not cause laws, or modality), one legitimately wonders about the kind of epistemic access we can have to this non-causal determinative relation. This will be another recurring theme in this book. As a matter of fact, my account is deflationary in another sense: I do *not* take constitutive explanation to be a real and objective matter. Thus, my variant of explanatory dispositionalism cannot be understood as an exploration of an objective explanatory structure of reality. There is no such thing. This point ties back to the aforementioned deflation of the distinction between derivative dispositions and fundamental powers: there are no such things.

Thus, I completely reject the naïve methodology according to which all it takes to establish explanatory dispositionalism is to put forward explanatory bridge-principles linking dispositional talk with nomic talk, or modal talk, or what have you. Finding these principles may indeed be part of the story, but it is not the entire story. An indirect methodology should be filling the gaps here. In this book I will submit that epistemically successful explanations in terms of dispositions ought to be considered as hypotheses, to be indirectly and increasingly (but also, crucially, defeasibly) corroborated by the theoretical successes of the theory they are embedded in. When a metaphysical theory, understood as a body of explanations, enjoys enough virtues over its rivals, then the sentences to be found at the body of these explanations have the more credentials to be thought as perspicuous. Such an evaluation of a theory, cannot, of course, completely be undertaken within the pages of this book – which justifies the brackets in the second sentence of this chapter. I cannot exhaustively prove, within the context of this book, that there is Dispositional Reality. I can clarify what Dispositional Reality is, and how should we proceed to justify our belief in it; I can even offer some partial motivation to believe in it. But none of this will be enough to convince the agnostic, or the openly sceptic. For this, we will need to wait for future development.

At this point, some reader might become worried. Here is a possible line of thought: by rejecting any objective explanatory structure of reality, how is your variant of explanatory dispositionalism a substantive metaphysical matter at all? Aren't you getting *too* deflationist for your own good?

As a matter of fact, I'm not. Or so I will argue. Matters of metaphysical perspicuity are objective, although we can only indirectly judge them through explanatory hypotheses. Dispositional Reality too is an objective matter, although whether dispositional talk can be taken to be perspicuous can only be indirectly gauged by assessing how well it fares as an explanatory primitive. To reassure the reader: I still believe metaphysics, and dispositionalism, to be a substantial and non-trivial matter about the nature of reality. As I will show, my rejection of a real and objective relation of constitutive explanation is compatible with that. Furthermore, it is not true that I reject any objective explanation. As I hinted before, I take explanatory dispositionalism to involve many different kinds of explanations, of which constitutive explanation is only one. Another explanatory notion clearly involved in the project is the peculiarly non-Humean notion of *governance*, viz. the idea that regularities in property distributions across space-time are not a brute and primitive matter, but happen because of something else. Governance and constitutive explanation, although both non-causal in character, are very different beasts; for one, unlike constitutive explanation, I take governance to be real, in the sense that it is underpinned by an out-there determinative relation, presumably of the non-causal variety (in this respect, I take governance to be more similar to causal explanation than constitutive explanation). The study of these various explanatory notions will be a throughline of the book.

This is the general sketch of the project. Of course, to perform such a radical rebooting, a lot of manoeuvring is required; indeed, to articulate such a novel and deflationary account of Dispositional Reality, our entire metaphysical set-up needs to be reconstructed from the ground up: from the analysis of metaphysical perspicuity, to the kind of philosophical methodology that accompanies it, and finally, to an in-depth study of the multiple types of explanations involved in my variant of explanatory dispositionalism, including governance.

For those of us interested in Dispositional Reality, there's a long road ahead. So, before we begin, it will perhaps help the reader to summarize the contents of the chapters in a somewhat organic fashion. This will complete the introductory chapter.

First of all, in the strictly interrelated Chaps. 2 and 3 I will articulate the metaphysical and methodological backbone for the rest of the work, which mostly revolves around the interplay of what I will call *reality* (of which Dispositional Reality is a part), and what I call *appearance*. This is only apparently a detour, and is crucial to understand what Dispositional Reality is. In Chap. 2 I will discuss the notions of metaphysical perspicuity and metaphysical structure. The idea of metaphysical perspicuity, as materially distinct from truth, is a crucial component of my approach, but it also suggests an inviting approach for metaphysicians like friends (and foes) of dispositions, allowing one to strike a useful middle ground between error-theory and meta-ontological deflationism. As a result, metaphysical debates such as the one about dispositions (but many other examples will be provided to contextualize these claims) should be understood as revolving around what is really the case, as opposed to what is merely the case. This is a popular re-orientation in contemporary metaphysics, but how ought it to be understood? I will claim that it ought to be understood as the claim that some sentences are not merely true, but metaphysically perspicuous, in the sense that they reveal metaphysical structure. Roughly put, there are two ways to understand these notions, usually associated to the work of two pairs of philosophers (as to whether this reconstruction is exegetically faithful, I won't say). These are the Russell-Wittgenstein Way to Structure and Perspicuity, and the Lewis-Sider Way to Structure and Perspicuity. I will argue against the former, and for the latter, which more closely mirrors some of the doubts expressed in this introduction, and which alone can deliver a suitably "non-ontic" notion of metaphysical structure.

In Chap. 3 I will deal with appearances. Firstly, I will argue for a factive conception of appearance (if it appears to be the case that ϕ, then ϕ), and against what I call a "lifting-the-veil" conception of metaphysics, according to which according to which metaphysics, as a theoretical inquiry, has the goal to re-evaluate and possibly discard ordinary and pre-theoretical beliefs. This supplements the contents of the previous chapter, and idea that error-theory is an undesirable alternative to the present approach. The argument I have in mind is metasemantic in nature. Of course, once we accept a factive conception of appearance, it remains to be seen what these appearances are: I argue that appearances should not be reified as shadowy or "less fundamental" separate parts of reality; thus I reject the idea that the popular distinction between what is really the case and what is only apparently the case ought to be understood as a grounding- or fundamentality-based layering of reality. We only need, *per* our account, the distinction between perspicuous truths and mere truths, a far less problematic distinction. No shadowy realm of mere appearance is needed, nor any layering aspect of reality bridging the more fundamental to the less fundamental. Finally, Chap. 3 is also where the notion of constitutive explanation, bridging reality with appearances, is first introduced. After arguing against its reality, my indirect method for metaphysics – loosely neo-Quinean in character – is proposed: although constitutive explanations are not real, reality still constrains our explanations, as certain bodies of explanations are inherently more virtuous than others.

With this background in mind, we can finally introduce Dispositional Reality, which is the topic of Chap. 4. The exceedingly impatient reader may very well skip to Chap. 4 if they are curious about Dispositional Reality, but should be ready to backtrack to Chaps. 2 and 3 to understand some of the most crucial concepts: for it is only within the context of my account of metaphysical perspicuity and structure, and my treatment of explanatory notions, that Dispositional Reality emerges as an appealing philosophical concept. Roughly put, to believe that there is Dispositional Reality is to believe that (at least some) dispositional talk is metaphysically perspicuous, and thus that it reveals a dispositional aspect of the structure of reality, to be understood along the non-ontic lines presented in Chap. 2. This dispositional aspect of reality, quite importantly, is not a thing, and thus not a property of any kind. The deflationary potential of the position is considerable; e.g., I will flirt with Ostrich Nominalism, understood as the position according to which simple predicative sentences are more perspicuous than sentences about propertyhood or instantiation. This is where my position emerges as a kind of power metaphysics without powers, equipped with many key virtues; firstly, it provides an ontologically inconspicuous way to reconstruct the distinction between metaphysically innocent and metaphysically loaded dispositional talk. Secondly, it sharply deflates many issues that have plagued dispositionalist metaphysicians in recent years, or at least reframes them in a much more manageable fashion; this goes to show that these may not be foundational issues, but rather the result of a faulty perspective.

In Chap. 5, I will approach the decisive question: what reasons do we have to think that there is Dispositional Reality at all? In doing so, we will need to pick up the methodology put forward in Chap. 3. Furthermore, this is where the inherently explanatory element of my position will come to the forefront. According to our loosely neo-Quinean methodology, to score a point for Dispositional Reality one must argue that dispositional talk explanatorily indispensable in our best possible accounting of the nomic family, which requires an indirect and holistic evaluation of metaphysical theories as bodies of explanations. An exhaustive argument to this conclusion, as one would expect, vastly outscopes the present book. Chapter 5 will also be an occasion to evaluate old dispositionalists arguments; my criticisms of the "naïve methodology" mentioned before, which entirely revolves around the adequacy of bridge-principles such as the conditional analysis, deflates much of the importance of the discussion concerning conditionals and interferences. I do not have much faith in the so-called "Argument from Science" either. As I will argue, to discover whether there is Dispositional Reality is a metaphysician's job, rather than a scientist's.

The topic of Chap. 6 will be the *extent* of Dispositional Reality. This might strike the reader as a kind of obscure topic, but it's actually quite simple. Even conceding that there is Dispositional Reality, is Dispositional Reality a homogeneous realm, or is some kind of dispositional pluralism to be preferred? Secondly, how much of reality is dispositional, viz. what other kind of non-overtly dispositional predication is ultimately underpinned by Dispositional Reality? As for the first question, I don't think it would be wise to "make a decision before its time", viz. to impose certain dispositional locutions as paradigmatic at the expenses of others; for as of now, given the multifaceted nature of dispositionality that we have to account for, there is little reason for a cogent choice. The way I see it, it all depends on which is best to perform the work: after all Dispositional Reality has been introduced as a working hypothesis. The choice between dispositional pluralism and its adversary, too, is an open question that depends on the explanatory successes of dispositionalism and the resources required to achieve it. In our pursuit of explanatory dispositionalism, I will help myself with whichever dispositional resource at my disposal.

The second question to be discussed in Chap. 6 amounts to the following: what other kind of non-dispositional predication is ultimately underpinned by Dispositional Reality? This is the distillate of the venerable choice between dispositional monism and dualism: are all properties dispositional, or only some of them? Although, within the context of my account, there is no metaphysical divide between dispositional and categorical properties, I will argue for what could reasonably still qualify as a form of dualism: Dispositional Reality is not a negligible region of metaphysical structure, but there is little reason to think that it exhausts it.

Chapter 7 is where finally some explanatory endeavours will be attempted. Primarily, I will explain how exactly dispositions (non-constitutively) explain instances of regularities, and how dispositions (constitutively) explain laws. Properly distinguishing between these two kinds of explanations is itself part of the job. Accounting for the non-constitutive explanation of regularity-instances means accounting for the notion of governance, which will be a protagonist of the chapter.

On the other hand, the explanation of laws crucially relies on a modal and explanatory correlation between dispositions and other properties (in a lightweight sense of the word) studied in the previous chapter. In line with the deflationist flavour of the rest of the book, dispositional *explanans* in both cases will simply consist in dispositional ascriptions, or some logical construct thereof – in contrast with, say, essence-based explanations such as the derivation in Bird (2007a, p. 46). The reason is simple: there may be inherent benefits in trying to exclude essence from our explanatory bases: as a smaller number of primitives is itself a theoretical virtue when it comes to comparing positions, within the dispositionalist camp and beyond, it may be worthwhile to develop my take on Dispositional Reality by cultivating the metaphysically deflationary flavour of the previous chapters. We want to prove that dispositions are part of the structure of reality: if we can do that without also having to prove that essences are a part of reality, that would be preferable. This chapter thus functions as partial corroboration for Dispositional Reality as an explanatory hypothesis.

Rather than continue with other explanatory endeavours, say counterfactuals or modality in general, Chap. 8 is devoted to the resolution of a problem which naturally emerges from the explanations in Chap. 7, and which is inevitably bound to pop up in any future development. I have claimed that, albeit in different ways, dispositions explain laws and regularities; but here is a question: how, exactly? What is it exactly about dispositions that make them suitable to function as *explanans*? Chapter 8 deals with these meta-explanatory questions, which I will investigate by borrowing on two recent debates, the first on the so-called "inference problems" for theories on properties and laws, and the second on meta-grounding. My solution is rather simple: my explanations are based on certain general principles, meta-explanatory hypotheses understood as postulational posits concerning the introduction of a metaphysically informed notion of dispositionality into the theory. Dispositionality is primitive on this account, in the sense that there is no reductive definition or constitutive explanation that will clarify, in simpler terms, what it is for something to be so-and-so disposed. There's no "dissecting dispositionality" to understand what it is about dispositions that make them suitable to account for the nomic or modal aspect of the world; explanations, after all, have to end somewhere. In Chap. 8, I will argue that dispositionality is a good place to stop. I will also argue against the possibility that such a meta-explanatory answer might be essence-based in character (a common strategy when it comes to meta-grounding), thus vindicating some form of essentialism. Not so on my account: essences, the way I see it, are explanatorily inert.

The resulting account is, interestingly enough, not trivial: these meta-explanatory links introducing dispositionality generate unique explanatory structures that are different from the one encountered in other competing theories, such as the so-called DTA theory of laws, dispositional essentialism, or even more remotely, neo-Humean metaphysical systems. Chapter 8 is, in this respect, crucial for the purpose of this book, for it shows that the beating heart of dispositionalism lies in its explanatory applications. Dispositionalism ought not to be understood as the introduction of a certain ontology, which can then, in a second logical moment, provide certain explanations. Dispositions is *inherently* explanatory.

Chapter 8 effectively concludes the bulk of the book. By presenting Dispositional Reality, setting up explanatory endeavours, and providing a systematic solution to meta-explanatory questions, the stage has been set for future research concerning Dispositional Reality. The goal of the book is, in other words, to present Dispositional Reality as a theoretically virtuous explanatory hypothesis of the non-Humean variety, a hypothesis which is not definitely settled but is increasingly corroborated by the successes of the theory, both present and future. Absolute certainty, unfortunately, is not of this world. The very brief and conclusive Chap. 9 will be an occasion to reflect on the results on the book and plan for future research.

Chapter 2
Reality

2.1 Introduction

To get a grip on the notion of Dispositional Reality, one must first understand what I mean with "dispositional", and what I mean with "reality". Let us start with the latter: in the next two chapters, I will articulate the meta-philosophical backdrop of my proposal, achieved by both characterizing the world surrounding us, and by offering some related methodological remarks as to how the philosopher should go on about exploring it: as we will see, much of this backdrop revolves around the interplay between what I call *reality* and what I call *appearance*. The key idea behind my approach is to deploy the notions of metaphysical structure and metaphysical perspicuity to draw the distinction between reality and appearances. This approach will be particularly fruitful when dispositions will be introduced. The impatient reader will perhaps wonder why I take so long a detour before explicitly dealing with dispositionality; but to see what is wrong in current debates in power metaphysics, we need to take a step back.

In this chapter, I will focus on the notion of reality, whereas in the next one I will deal with appearances; given the close interconnections between these topics, there is bound to be some back and forth between the two chapters, but I will try to keep things as straightforward as possible.

L. Azzano, *Dispositional Reality*, Synthese Library 482,
https://doi.org/10.1007/978-3-031-52625-1_2

2.2 Metaphysical Realism, Structure, and Perspicuity

My starting point is Metaphysical Realism: viz., there is a reality which does not depend on our linguistic or conceptual activities; relatedly, it is the job of the metaphysician (perhaps not entirely their job, nor exclusively their job) to investigate it. I will not argue for this claim.[1]

So understood, Metaphysical Realism is a vague, almost nondescript claim: one may try to make it more specific by claiming that, e.g. "[m]etaphysical realism is the thesis that the objects, properties and relations the world contains exist independently of our thoughts about them or our perceptions of them" (Khlentzos, 2021), or perhaps by claiming that there is an external world of mind-independent facts; after all, if moral realism is the claim that there are mind-independent moral facts, metaphysical realism might consist in the claim that there are mind-independent facts *tout court*. I do not accept these suggestions: for whether reality consists in objects, properties, or relations, is another matter entirely – and one which I will not settle in this book; similarly, I will only claim that there are mind-independent facts under the most deflated understanding of "fact".

Although I do not feel particularly committed to the claim that reality is made up by objects, properties, or facts, I do not conceive it, to use a famous metaphor, as an amorphous lump.[2] There is, on the contrary, *metaphysical structure*, which may be revealed by certain fragments of language, e.g. sentences. For a comparison, let us consider the position which Rayo (2013, p. 4) calls Metaphysicalism. Metaphysicalism is the conjunction of two claims, the first metaphysical and the second semantic. In general, (Rayo, 2013, p. 6–7)

- Metaphysicalism^{M}: "there is a 'metaphysically privileged' way of carving up reality into its constituent parts."
- Metaphysicalism^{S}: "in order for an atomic sentence to be true, there needs to be a certain kind of correspondence between the logical form of a sentence [...] and the metaphysical structure of reality."

When combined, we obtain Metaphysicalism, viz. a position according to which "in order for an atomic sentence to be true, there needs to be a certain kind of correspondence between the logical form of the sentence and the 'metaphysical structure' of reality'" (Rayo, 2013, p. 4).

These claims might sound mysterious to the uninitiated: what are these "parts" reality is "carved up into"? And how should we understand this "correspondence" between them and the logical form a sentence? I will answer these questions in due

[1]On this matter, I agree with the non-argumentative and almost biographical passage from Sider (2011, p. 18): "[t]his picture is perhaps my deepest philosophical conviction. I've never questioned it; giving it up would require a reboot too extreme to contemplate; and I have no idea how I'd try to convince someone who didn't share it."

[2]E.g., Hirsch (2008) and Eklund (2008); according to Eklund, the expression traces back to Dummett (1981, p. 577).

time; for the moment, Metaphysicalism^{M} should simply be understood as the claim that there is metaphysical structure, and Metaphysicalism^{S} as the claim that true (atomic) sentences reveal such structure. Similarly, the idea of "revealing structure" sounds just as mysterious as structure itself. Allow me to label the problem, instead of solving it: when a true sentence reveals, rather than obfuscate, some portion of metaphysical structure, let us say that the sentence is *metaphysically perspicuous*, or *perspicuous* for short. Metaphysicalism^{S} is then the claim that all true (atomic) sentences are perspicuous.

According to Rayo (2013, p. 10ff), Metaphysicalism is "bad philosophy of language", and his reasons are, in my opinion, quite convincing. However we desire to flesh out the notions of metaphysical structure and perspicuity, the fact of the matter is that fluent speakers of a natural language such as English do not appear to be interested in conveying information about structure when speaking; to borrow Rayo's example, the choice between interchangeable expressions such as "someone died" and "a death has occurred" is made for a variety of different reasons, but respect for structure does not appear to be one of them. The contrary opinion, that fluent speakers of English unknowingly and systematically utter falsehoods whenever engaging in non-perspicuous talk, seems too extreme of an error-theory to be taken seriously – especially because there is little indication that these speakers are as incompetent as the error-theory would make them out to be. I will come back to this point multiple times throughout Chaps. 2 and 3; for the moment it is important to notice that the culprit for this *prima facie* flaw of Metaphysicalism is the semantic component, and not the metaphysical one; viz. the problem is not that there is metaphysical structure, but that the only truths are those that reveal it. Rayo himself, although not particularly sympathetic to the notion of metaphysical structure, states that the real culprit here is Metaphysicalism^{S}, as a thesis on the relation between truth and perspicuity. Tentatively, let us discard Metaphysicalism^{S}, but keep Metaphysicalism^{M}. There is metaphysical structure, and some perspicuous truths reveal it; but not all of them: there are non-perspicuous truths as well. A true but not perspicuous sentence, I will call a *mere truth*.

So, truth and perspicuity are not the same thing, not even materially. Relatedly, one may wonder about the other direction of the biconditional linking truth with perspicuity: are all perspicuous sentences true? Answering yes to this question amounts to the adoption of what I will call Factivity of Perspicuity: according to it, if a sentence is perspicuous, then it is true; in other words, to say something revealing about the structure of reality, one *at least* has to say something true. As in the case of Metaphysical Realism, there's very little that I can say in favour of this principle; given the fact that I intend to use the notion of structure and perspicuity to frame the distinction between reality and appearance, rejecting the Factivity of Perspicuity would be at best peculiar, and not something to be done without independent motivation; for it would amount to the claim that faithful representations of the make-up of reality can be achieved by uttering falsehoods (Lipman, 2018, p. 592). Thus, I will proceed under the assumption that, although there can be non-perspicuous truths (viz. mere truths), there is no such thing as non-true perspicuity: mere truths are truths.

2.3 Why Metaphysical Structure? And Why Non-perspicuous Truths?

At this point I should probably motivate my choices: why accept Metaphysicalism^{M}? And why reject Metaphysicalism^{S}? We have seen some *prima facie* justifications for the rejection of Metaphysicalism^{S}, but their philosophical weight is dubious, especially if we do not find some motivation for Metaphysicalism^{M}, viz. for enriching reality with structure in the first place. These questions are made the more pressing by the fact that Metaphysical Realism is not *per se* enough to support Metaphysicalism^{M}: it is one thing to say that there is a mind-independent reality, and another thing entirely to say that reality displays structure.[3] So, even conceding Metaphysical Realism as a background assumption, more needs to be said about the existence of metaphysical structure and the divarication between truth and perspicuity.

I do not have a direct argument in favour of the existence of metaphysical structure, nor in favour of the existence of mere truths. What I have is, rather, a very indirect methodological point: that to accept Metaphysicalism^{M} while rejecting Metaphysicalism^{S} makes for a very appealing meta-metaphysical set-up. When properly understood (more on that below), we can use the distinction between perspicuous sentences and mere truths as a way to set up the divide between reality and appearances, and thus solve many familiar puzzles concerning the compatibility between ordinary sentences and certain metaphysical positions (this will also make for a compelling starting point for the treatment of dispositions). Finally, after having motivated their introduction, I will analyze the notions of metaphysical structure and perspicuity, which will conclude this chapter.

2.3.1 Moorean Arguments

I mentioned "familiar puzzles". With that, I meant a family of well-known arguments in meta-metaphysics, which I will refer to with the umbrella term of "Moorean Arguments". Moorean Arguments are, supposedly, a variegate family, in both form and content, but they all have some things in common: they start with what appear to be obviously true premises, occasionally called "Moorean certainties".[4] These are things which we "ordinarily believe" (Fine, 2001, p. 2), and they supposedly constitute pieces of "everyday knowledge" (Lewis, 1996, p. 549) which are "more

[3] See Rayo (2013, p. 24–25). More generally, the possibility of "realism" without structure, or at least without a privileged description of reality such as the one afforded by metaphysical perspicuity, appears to be a common goal amongst deflationists or "easy" ontologists; e.g., Hirsch (2002).

[4] While the term "Moorean Argument" is somewhat artificial, the notion of a Moorean certainty is more widespread in the literature –as we will shortly see. The nomenclature stems from E.G. Moore's famous argument against skepticism, based on the premise "here is one hand"; whether this argument fits the mold of Moorean Arguments as I will discuss them, is unclear. I will come back to this exegetical point in Chap. 3, when I will discuss the source of this supposed "Moorean certainty" for the premises of such arguments.

credible than any philosopher's argument to the contrary" (Schaffer, 2009, p. 357); one might wonder whether such unassailable knowledge is so readily available, but let us grant this point for now. In a Moorean Argument, such premises are used to infer a conclusion which directly contradicts a certain metaphysical thesis (usually a position of the anti-realist variety). The "puzzle" here, if any, is not constituted by the fact that there is an argument against that position, but that such an argument requires so little to function – viz. only the Moorean certainty in the premise. Thus, the debate surrounding that position can be shown to be not a profound matter of metaphysics, but something which can be established on extra-metaphysical, almost trivial grounds.

The simplest Moorean Argument is one concerning the ontology of arithmetic. Suppose that you are a mathematical nominalist. Surely the following sentence is true:

(1) There is a prime number between 3 and 7

From (1) one can infer

(2) There are numbers

(Let us assume, for the sake of the argument, that the inference from (1) to (2) is valid; there are various ways to reconstruct this step in both natural and formal language, but there is little utility here in harping on details). Unfortunately (2) seems to be in contrast with the nominalist's position: if the nominalist position is understood to claim or entail that there are no numbers, (2) is its negation.

What are the options for the mathematical nominalist? The first option, (A), simply consists in waving a white flag: to accept (1), the inference from (1) to (2), and thus accept (2) as well, and, in conclusion, to reject nominalism. This means that nominalism is open to almost-trivial counterexamples; given how widespread Moorean Arguments are, this kind of approach entails that many metaphysical theses of these anti-realist variety are not worth entertaining. Here is another case: according to mereological nihilists there are no mereological fusions, viz. things with proper parts; such claim can easily be threatened by a Moorean Argument starting with the premise "there is a tree in Lorenzo's backyard", and if option (A) is chosen, such an inconspicuous sentence is all it takes to falsify their position.[5] Similarly, according to Fine (2001), moral anti-realism, understood as the position according to which there are no moral facts, could be easily falsified starting from the ordinary premise "killing babies for fun is wrong".[6]

[5]Unless one finds a way to reconcile the truth of "there are trees" with the truth of "there are no mereological fusions", as in Contessa (2014). This anticipates another way to deal with Moorean Arguments that I will discuss below.

[6]Although Fine does not explicitly recognize it, the Moorean Argument against moral anti-realism is a little bit more complicated; only under the schematic assumption that "if p, then there is the fact that p" that "killing babies for fun" can falsify moral anti-realism, understood as the position according to which there are moral facts. The peculiarities of the Moorean Argument against moral anti-realism will become of crucial importance shortly.

In general, option (A) amounts to a deflationist attitude towards ontological or metaphysical disputes, according to which such disputes can easily be resolved on extra-metaphysical grounds – such as basic arithmetic proficiency, or a simple empirical investigation. If one wants to avoid option (A), there is option (B), which consists in rejecting (1) instead. (B) is the *modus tollens* to (A)'s *modus ponens*. Namely, according to option (A): if there is a prime number between 3 and 7, then mathematical nominalism is false; there is a prime number between 3 and 7. So: mathematical nominalism is false. But according to option (B): if there is a prime number between 3 and 7, then mathematical nominalism is false; mathematical nominalism is not false. So: there is no prime number between 3 and 7.

According to option (B), sentences such as (1) may be advanced as "loose talk", or in a make-belief spirit, but they are literally false; thus, according to the mathematical nominalist who chooses (B), there is no prime number between 3 and 7. This conclusion may strike the reader as so absurd that it may very well backfire and cement (1) as a Moorean certainty; but as radical as it is, (B) is not without its supporters (Field, 1980, 1989; Yablo, 2001). More generally, option (B) shields our metaphysical theses at the price of an ever-pervasive error theory, spreading well-beyond the boundaries of philosophical discourse. The problems plaguing (B) as a kind of error-theory are multiple, from the apparent inability to distinguish between, say, "good" arithmetic talk from "bad" arithmetic talk, to the inability to properly engage with mathematical discourse without contradicting oneself; furthermore, one may harbour doubts as to whether the premises of Moorean Arguments, *qua* Moorean certainties, can be false in the first place.[7]

So, if one intends to navigate between the Scylla of deflationism (A) and the Charybdis of error-theory (B), perhaps a third option would be preferable. Enters option (C) – where "C" stands for "compatibilism". As different as they are, (A) and (B) agree on one thing: on the truth of the conditional "if there is a prime number between 3 and 7, then mathematical nominalism is false". (C) disagrees. Loosely put, the goal of (C), as a kind of anti-realism without error-theory, is to preserve the relevant form of anti-realism as a substantial philosophical position which cannot be dismissed by quasi-trivial observations such as "there is a prime number between 3 and 7", or "there is a tree in Lorenzo's backyard". Given that so many of such sentences take the form of simple existential quantifications, the idea behind compatibilism is that many of the relevant forms of anti-realisms will need to be formulated in some other terms (viz., mathematical nominalism will not simply amount to the claim that there are no numbers).

[7] I will come back to Moorean certainties in the next chapter, when I will discuss the idea of a world of appearance being described by true sentences.

2.3.2 *Varieties of Compatibilism*

As far as I know, there are two broad families of positions within the compatibilist camp. According to the first, the anti-realist employs a different reading of the quantifier than the one used for the interpretation of the premise of the Moorean Argument; this strategy can either be pursued in a deflationist spirit (Hirsch's "quantifier variance") or in a non-deflationist spirit (Sider's "Ontologese gambit"). To the extent in which languages are individuated by their interpretation, this amounts to the claim that, say, the mereological nihilist is speaking another language, albeit one very similar to English ("Nihilese", or perhaps "Ontologese", the special language of the ontology room). I will not spend much time discussing these options; this is because, long story short, I am not very confident that these languages really exist. Two quantifiers (or quantifier-meanings) are different if they can be *shown* to be different – just as, say, classical negation can be shown to be different from intuitionistic negation because the former, unlike the latter, satisfies double negation elimination. But where is the difference between the quantifiers of English and Nihilese, or Ontologese? To the extent in which they are all understood unrestrictedly, and their inferential behaviour is that of an ordinary quantifier (viz., they satisfy standard rules of introduction and elimination), then there is very little to differentiate them, beside the wishful thinking of the compatibilist.[8]

Whereas this first family of compatibilist positions focuses on quantifiers, the second one focuses on whole sentences. Borrowing from an approach in Fine (2001) which has proven itself to be exceedingly popular in recent literature, we might distinguish between what is really the case, and what is merely the case.[9] Fine's approach is ultimately the source of what has been defined a "significant reorientation [...] in analytic metaphysics" (Koslicki, 2012, p. 186) according to which metaphysics is not primarily about what there is, but about what is really the case. According to such a re-orientation, questions of

---

[8]This kind of criticism has been leveraged against Hirsch's quantifier variance as a "collapse objection" –in the sense that if the two quantifiers are inferentially identical to the point of logical equivalence, the compatibilist strategy is bound to fail. See Hale and Wright (2009), Eklund (2009), Dorr (2014); for a rebuttal, see Warren (2015). Recently Bueno and Finn (2018) have also questioned the lack of difference between the quantifiers of English and, say, Nihilese. The case of Ontologese is slightly different, as its quantifiers are allegedly stipulated to be joint-carving (Sider, 2011, p. 204); it does not matter what "joint-carving" means in this context; for the strategy to pay off, the two quantifiers must *not* be logically equivalent, which they are if they are inferentially identical –as Sider *(ibid.)* admits they are.

[9]Fine (2001, p. 3) prefers the expression "only apparently the case", but I take this expression to have unfortunate nuances; to state that something might only apparently be the case suggests that it might *not* be the case. This seems to go hand-in-hand with a perceptive reading of appearances that would support the possibility that what is only apparently the case is not the case (viz., it is only apparently the case that the stick immersed in water is bent, but it is not the case). But this is not the goal here: whereas Fine's goal, as is ours, is to avoid error-theory, and thus ensure that what is only apparently the case and what is really the case are equally the case. Thus, I prefer the expression "merely the case", which more transparently conveys this. In the next chapter I will discuss more in detail the factivity of appearances, and argue in favour of it.

realism can be suitably recast; thus, e.g. to be a realist about Ps is not simply to claim that there are Ps, but that it is *really* the case that there are Ps. This provides a form of compatibilism, in the sense that once the mathematical nominalist's position is taken to state (or entail) that there really are numbers, (2) does not contradict it anymore. But whence the difference between there being numbers and there really being numbers? The qualification might strike as little more than rhetorical foot-stomping, and, as Fine himself recognizes, some elbow grease needs to be put into this claim to make it intelligible: "[f]or what room is there, in our ordinary conception of reality, for any further distinction between what is genuinely a fact and merely the semblance of a fact?" (Fine, 2001, p. 8).

The simplest clarification of this strategy, as in Schaffer (2009), is one which I want to avoid: that of claiming that reality, possibly but not necessarily a world of facts, displays a layered structured imposed by an ordering relation of *metaphysical grounding* or *relative fundamentality*, ranging from the least fundamental to the more fundamental; if this relation is well-funded, this is where the interest of the philosopher will lie. This is a bit too much for my taste. I have claimed before that although I am a metaphysical realist, I do not assume that this objective mind-independent reality is parsed into facts and/or objects; so surely I am not going to take on board the even more controversial assumption that there are (more or less) fundamental facts and/or objects. The problem, mind you, is not only that this strategy is metaphysically cumbersome, but also potentially mysterious: it does not simply populate reality with a multitude of facts and/or objects, but more or less fundamental ones; this is what we may call a layered conception of reality.[10]

A less cumbersome version of this strategy, which makes do without resort to a layered conception of reality, deploys truthmaking instead (Cameron, 2008; Heil, 2016). According to this strategy, say, the mereological nihilist is not threatened by the truth of "there is a tree in Lorenzo's backyard", since what matters for the metaphysician is what makes it true, viz. its truthmaker: for the nihilist, a collection of mereological atoms arranged tree-wise. Cameron (2008, p. 7) articulates the virtues of this proposal in the clearest terms:

> Do not think of the distinction as dividing the entities in the world into the privileged real existents and the impoverished unreal existents. All there is in the world are the real existents: once you've said what there really is, you have answered the ontological question. The distinction between mere existence and real existence is just a way of talking. The rules of the language are that 'a really (or fundamentally) exists' is true iff a is an element of our ontology (read: iff a does some truthmaking); that 'a exists' is true iff 'a exists' is made true by some thing(s); and that 'a merely (or derivatively) exists' is true iff 'a exists' is made true but isn't made true by a. Stripped of any worrying metaphysical overtones, the distinction between real and mere existence (or, equivalently, fundamental and derivative existence) should not cause us unease.

[10]There are deflated conceptions of grounding which do not require the reification of what is grounded and what is doing the grounding; usually they function by regimenting grounding not as a predicate but as a sentential operator, e.g., Correia (2010), Fine (2012), and Dasgupta (2017). In the next chapter I will deal more explicitly with the nature of appearances and their relation with reality, and in that context, I will put submit that even such deflated conceptions require some degree of problematic layering.

Fine's potentially problematic distinction between what is really the case and what is merely the case can be similarly neutralized, e.g. by taking truthmakers to be facts: it is really the case that p if and only if "p" is true and it is made true by the fact that p; whereas it is merely the case that p if and only if "p" is true but it is not made true by the fact that p. However, truthmaking accounts still require either objects or facts to populate reality to function as truthmakers – which is again something that I do not wish to assume.[11]

2.3.3 Compatibilism Through Structure

Is there some other way to understand Fine's distinction, to obtain a new compatibilist option (C)? This is finally where the acceptance of Metaphysicalism^M and the rejection of Metaphysicalism^S can earn their keep (Sider, 2014); as I reminder, the idea is that reality displays some degree of metaphysical structure, yet sentences can be true without being metaphysically perspicuous (they obfuscate, rather than reveal, structure).

This naturally provides a way to clarify Fine's distinction between what is really the case and what is merely the case. Schematically,

F_1) It is really the case that p if and only if "p" is metaphysically perspicuous
F_2) It is merely the case that p if and only if "p" is merely true

Given the Factivity of Perspicuity and a simple principle of truth-disquotation, if it is really the case that p, then p. However, the other side does not hold; viz., it is not the case that if p, then it is really the case that p; this is because some truths are mere truths.

Loosely put, for a metaphysician of Finean sympathies, Moorean Arguments present themselves as a conflict between what is really the case and what is merely the case. E.g., it may appear to be the case that there is a prime number between 3 and 7, but if the mathematical nominalist is correct, that is not really the case. According to the error-theory of (B), if this conflict is genuine, it simply amounts to the distinction between true and false sentences; but on this approach, the conflict between reality and appearances should be understood through the distinction between perspicuous sentences and mere truths. So the mathematical nominalist should accept the truth of (1) and (2) ("there is a prime number between 3 and 7", and "there are numbers"), but deny that the sentence is metaphysically perspicuous. Similarly, the mereological nihilist will accept that "there are trees" is true, but deny that it is perspicuous; and similarly, for the moral anti-realist for the sentence "killing

[11] Just as there is a deflated conception of grounding, there is a deflated conception of truthmaking, that by regimenting the notion as a sentential operator, does not require the postulation of truthmaking entities. See Hornsby (2005), Melia (2005), Schnieder (2006), and Mulligan (2007). As for the case of deflated grounding, I will come back to this in the next chapter.

babies for fun is wrong". Various forms of realism and anti-realism, on this account, are understood as position concerning which sentences are metaphysically perspicuous, and which ones are not. Moorean Arguments, as valid as they are, cannot get any traction against the metaphysical positions so formulated, because it is nowhere being argued that the relevant sentences are not just true, but perspicuous.

One may object that this reconstruct entirely misses the point of these positions; e.g. moral anti-realism is, after all, supposed to be a position about morality and not sentences, their truth or their perspicuity. But remember what we have said about metaphysical perspicuity; to say that a sentence is perspicuous is to say that it matches, or reveals, some portion of metaphysical structure; so, ultimately, to claim that "killing babies for fun is wrong" is not perspicuous, is to claim that it is not revealing any moral aspect of objective reality – whether that will take the form of moral properties, facts, or what have you. And similarly for all the metaphysical positions targeted by Moorean Arguments: to make a claim about the perspicuity of a sentence, is ultimately to make a claim about reality.

Some of the advantages of this strategy are evident. Firstly, it avoids problems commonly associated with error-theory. Secondly, it does not involve any trickery concerning quantifiers and interpretations thereof, nor does it stipulate a hypothetical language for the metaphysician to speak. Thirdly, it does not impose a layered conception of reality: and it does not require, at least at this stage of development, any metaphysical set-up concerning objects, properties, facts or what have you. There is no metaphysically diminished "realm of appearances" corresponding to Moorean certainties: there only are non-perspicuous truths, perspicuous truths, and the structure they reveal.

Fourthly, and lastly, this approach allows us to recast Moorean arguments under a new light; more specifically, it allows us to better understand the feeling of miscommunication that we might have concerning certain exchanges. E.g., consider the following exchange:

Layperson:	"is there a prime number between 3 and 7?"
Mathematical Nominalist:	"I don't think that in reality there are such things as numbers"

Something has gone wrong in this exchange, clearly. The two parties appear to be miscommunicating somehow. The layperson asked a question on what seemed to be a clearly defined topic – and, mind you, a question with a clear answer for anyone with basic arithmetic proficiency, and the mathematical nominalist has provided what appears to be an irrelevant and unrelated answer. It is not far-fetched to imagine the layperson follow this answer by reiterating the question. The same occurs for other cases:

Layperson:	"do you think that killing babies for fun is wrong?"
Moral Anti-Realist:	"actually, I don't think that morality is real"

The layperson might yet again find this answer unsatisfactory, if not outright suspicious (imagine someone uttering "I don't think that morality is real" during their trial for murder). What is going on, here?

One idea might be that the feeling of miscommunication is due to the employment of different quantifiers and interpretations thereof. I have already expressed my misgivings towards this idea. My strategy for achieving compatibilism may suggest a different reading of the situation, one that has to do with shifting interests in certain contexts of utterances. The context of utterances is important for a number of reasons; for our purposes, it is important in the following way: context decides whether one is interested in the truth or in the perspicuity of the corresponding sentences. In certain contexts, we utter "p" and simply imply that "p" is true; in other words, we utter "p" for matters of factuality, to assert what is the case; in other cases we utter "p" and also imply that "p" is metaphysically perspicuous. These are the contexts in which we talk with an interest in the structure and make-up of reality, and thus in these contexts we will probably find some explicitly theoretical contexts; e.g. the ontology room. For simplicity, let us call the former contexts Ordinary Contexts, and the latter contexts, Theoretical Contexts. Because of the distinction between them, sometimes we may be interested in factuality without being interested in structure. E.g.,: in a primary school setting, someone with nominalist tendencies will have no trouble uttering, e.g. "there are prime numbers"; but within the context of, say, a discussion on a book on fundamental metaphysics, they will probably shy away from such an utterance. This does not require any degree of inconsistency from them: it is merely a consequence of their shifting interests, first in truth and then in perspicuity.[12]

Given this background, we can see what is going on these exchanges above. The layperson's utterance was done within an Ordinary Context, in which the layperson was simply interested in matters of fact: whether there is in fact a prime number between 3 and 7; perhaps the layperson is a teacher, who is simply interested in

---

[12] Here is an interesting thing to notice; the mathematical nominalist should probably refuse to utter "there are prime numbers" in a Theoretical Context; not because they do not believe that there are prime numbers; but because they do not accept the perspicuity of that sentence. But how should they express their disagreement with someone who does utter "there are prime numbers" in such a context? They may probably want to refrain from uttering "there are no prime numbers"; the mathematical nominalist of this persuasion after all accepts the truth of sentences "there are prime numbers", they just reject their perspicuity. This is why linguistic particles like "actually" or "really" or "in reality" are added; to express dissatisfaction with the supposed perspicuity of a sentence without flat-out denying it. Thus Mathematical Nominalist, when they perceive to be in a Theoretical Context, may refuse to utter "there are no prime numbers", or utter something along the lines of "in reality, there are no prime numbers"). Incidentally, I agree with Fine (2009, p. 168) that it is unfortunate that realist and anti-realist positions are usually formulated through expressions such as "there is...", which so easily lend themselves to metaphysically lightweight readings. In my account, this is explicated as follows: "there is..." is a very common expression in almost all linguistic contexts, whereas other expressions such as "...is real" more pointedly suggest a context of utterance associated with theoretical investigation (whether in the ontology room or not), in which perspicuity matters more than truth; it may be a Moorean certainty that there is a prime number between 3 and 7, but it is surely not a Moorean certainty the prime number between 3 and 7 is real –and it is presumably impossible to formulate a Moorean Argument to that conclusion. In what follows, however, I will prefer to directly talk in terms of perspicuity and structure, rather than deploying expression such as "...is real" or "in reality...".

testing one's arithmetic proficiency; at any rate, the context is extra-metaphysical. Yet, for whatever reason, the mathematical nominalist mistakes this as a Theoretical Context, and thus refuses to answer "yes, there is a prime number between 3 and 7" because of the mistaken impression that, in the current context, that utterance will be taken to imply not only that there are numbers, but that "there is a prime number between 3 and 7" is metaphysically perspicuous, which they deny. Thus, they may utter a qualified sentences such as "I don't think that in reality there are numbers". This answer only indicates their belief in the claim that "there are numbers" is not metaphysically perspicuous; but it does not settle the truth of the original sentence "there is a prime number between 3 and 7". Whether "there are numbers" is perspicuous does not settle whether it is true that there is a prime number between 3 and 7. And so, the original question of the layperson, made within an Ordinary Context, is not answered. Thus, the miscommunication. The layperson wanted to know about whether there is a prime number between 3 and 7, not about whether in reality there are numbers – in other words, they were interested in matters of factuality, but not structure.[13]

2.3.4 Lipman's Problem

There is at least one problem for this strategy – albeit one that, in my opinion, is eminently manageable.

To see the difficulty, consider that, once a mathematical nominalist accepts Fine's distinction, an ambiguity in the positioning of the negation immediately appears. Two versions of the position are available:

Mathematical nominalism (1):
It is really the case that: ~(there are numbers)

Mathematical nominalism (2):
~(It is really the case that: there are numbers)

The two are not equivalent, and the first option is in fact unacceptable. Remember that, by the Factivity of Perspicuity, if it is really the case that p, then p. But then mathematical nominalism (1) has disastrous consequences: for if the mathematical nominalist is committed to the claim that it is really the case that there are no numbers, then they are also committed to the claim that there are no numbers.

[13] My stance on this matter is interestingly related to the one in Yablo (2014); Yablo's stance, in not so many words, is that such miscommunications betray a perceived difference in aboutness of sentences in context. The difference between our accounts would be, that to the extent in which aboutness is understood as a component of meaning, that would still amount to some kind of semantic miscommunication, whereas I have claimed that these troubles do not have their source in the two parties' stance on truth, and thus presumably semantics of it, but on their attitudes towards perspicuity. I will come back to truth, and thus semantics, in the next chapter.

This, unfortunately, would make our entire compatibilist strategy for the mathematical nominalist ineffectual, because then the position would be endangered by the Moorean certainty "there is a prime number between 3 and 7". The Moorean Argument would still disprove mathematical nominalism, and we would be back to square one. The mathematical nominalist who has followed our suggestions so far should instead claim (2); in other words, they should not claim that "there are no numbers" is perspicuous, but that "there are numbers" is not. A nominalist of this stripe could claim that mathematical nominalist is an anti-realist position; it is not a position about how reality is, but about how reality is not; more specifically, reality is not numerical (whether this attitude is in fact shared by all nominalists is highly dubious, but never mind that). At any rate, the Factivity of Perspicuity generates further constraints about realists and anti-realists position framed in terms of metaphysical perspicuity. More specifically, no metaphysician is in a position to claim, given a sentence p, that p is perspicuous, if p is not consistent with what is merely the case, viz. with the mere truths from which we draw Moorean certainties, and thus the premises of our Moorean Arguments. In the words of Lipman (2018, p. 8), who clearly expressed this difficulty: "[t]he metaphysician cannot believe in the [...] reality of anything that conflicts with the ordinary facts". This is presented as a problem. But is it really? Given the aforementioned strategy concerning the positioning of negation, this is not really a problem for any of the anti-realist positions seen so far: such positions do not make positive claims about which sentences are perspicuous, but negative claims about which sentences are not. The problem is indeed pressing for any realist position which claims that some sentences are perspicuous, if it is indeed the case that the negation of such sentences belong to the special class of Moorean certainties. Is there some available example?

Maybe there is one. In mereology, the issue of compatibilism is discussed not only with respect to mereological nihilism, but mereological universalism; unlike mereological nihilism, mereological univeralism is not a form of anti-realism: on the contrary, it claims that for everythings (plural) there is something (singular) which they compose. But just as mereological nihilism might be threatened by sentences like "there is a tree in Lorenzo's backyard" or "there are trees", mereological universalism might be threatened by sentences like "there are no cat-dogs" (a cat-dog being the mereological fusion of a cat and a dog). The strategy I employ, however, allows for no compatibility here. The compatibilist could say that "there are no cat-dogs" (like "there are trees") is true but not perspicuous – but that would still be incompatible with mereological universalism. In short, it seems to me that the distinction between truth and perspicuity does not shield realist positions from Moorean Arguments which start from negative existentials. Thus, there is an asymmetry in the applicability of our Strategy (C) for Moorean Arguments: it can be applied to salvage anti-realist positions, but not realist ones.

My solution to this difficulty is to bite the bullet. This asymmetry of compatibilism is to be taken as a feature, rather than a bug. I will begin by treating the example of the Moorean Argument against the mereological universalist, not because I intend to piecemeal a solution, but because I take it to be very illuminating.

Consider this: is "there are no cat-dogs" really a Moorean certainty as "there is a prime number between 3 and 7"? I think not. We will come back in the next chapter to the unassailability of Moorean certainties (and we will then see why exactly "there are no cat-dogs" is not one of them). But we can already see that something is off here. Unlike some of the other sentences seen before, "there are no cat-dogs" displays a technical term "cat-dog" whose meaning is settled in the ontology room; a layperson may surely be instructed as to its meaning, and may then utter "there are no cat-dogs", but by that point they are not offering a pre-theoretical observation, or a piece of "everyday knowledge" (Lewis, 1996, p. 549): they are rather doing metaphysics, albeit very naively. Similarly, Doulas (2021) characterizes Moorean certainties as those things that "we ordinarily believe": I'm not sure who "we" is supposed to be, but I am going to go out on a limb and say that we do not ordinarily entertain the belief that there are no cat-dogs. "There are no cat-dogs" is a sentence whose components are embedded in a very specific formal theory, and it is thus unclear why it should enjoy any privileged status over other sentences of that theory. If there are Moorean certainties in the form of negative existentials, they should probably be limited to more unthreatening things like "there is no king of France", or "there is no Pegasus". So, if there are realist positions for which my strategy cannot achieve compatibility, then they will be, e.g. realism about the king of France, or realism about Pegasus. I will happily leave them to their troubles.[14]

To cement this point further, consider the following asymmetry: there is a feeling of miscommunication between the layperson and the mereological nihilist, which is *not* present between the layperson and the mereological universalist. The layperson claims "there is a tree in Lorenzo's backyard"; perhaps they are interested in building a treehouse, or finding shade on a hot summer afternoon, and in these Ordinary Contexts, truth and factuality matter more than metaphysical perspicuity; in such contexts, the mereological nihilist's reply "in reality, there are no trees" is a bad reply. Something has gone wrong in the exchange: the layperson was interested in the truth of some tree-talk, while the nihilist was interested in its perspicuity. But now consider the layperson claiming "there are no cat-dogs", and the mereological universalist replying "in reality, there are many cat-dogs". No miscommunication is perceived here. By virtue of the use of a technical term "cat-dogs", there doesn't seem to be any non-Theoretical Context of utterance available, which does not privilege metaphysical perspicuity over mere truth. Layperson or not, anyone uttering "there are no cat-dogs" is doing metaphysics. No compatibility is possible, nor desirable, between this claim and the mereological universalist's. They will have to discuss.

---

[14] Is it not possible to find philosophically problematic Moorean certainties in the form of negative existentials? The best candidates for the formulation of such sentences would be those expressions used in philosophy which are not exclusive purview of the philosopher, such as "person/people", "mind", "time", etc... But no Moorean certainty in the form of negative existentials can be easily reached through those; on the contrary, "there were 2200 people on the Titanic" sounds like a positive Moorean certainty to me, as is "there was a time when Europe was at war".

Generalizing these observations, we can sketch an answer to Lipman's problem. Anti-realist positions have it easy: by denying the perspicuity of certain true sentence, they can ground their claim that some appearance does not correspond to reality (so to speak). The case of the realist is different: in many cases, realism about Ps is not in conflict with any aspect of the world of appearance, simply because there is no Moorean certainty "there are no Ps"; e.g. there is no Moorean Argument against scientific realism, because there is no Moorean certainty "there are no electrons"; it's not that it is obvious in our everyday life that there are no electrons – we just lack certainty, one way or the other. It is however certain that there is no king of France; hence, there might be a genuinely insoluble conflict between appearance and reality for, say, someone who endorses realism about the king of France.

This is a fascinating topic, which vastly outstrips our current purpose; let us concentrate on Moorean Arguments against anti-realist positions such as mathematical nominalism, mereological nihilism, moral anti-realism, and so forth; in these cases, my compatibilist strategy based on metaphysical structure and non-perspicuous truths constitutes an appealing option, surely less cumbersome than many of its competitors in the compatibilist market– thus offering some *prima facie* motivation for the adoption of this kind of framework. As we will see in Chap. 4, there are Moorean Arguments against power anti-realism as well, which are easily treatable in the same fashion: this will properly kickstart our discussion on Dispositional Reality.

2.4 The Ontological Temptation

Before tackling the nature of structure and perspicuity, there is one last point – of crucial importance for the purposes of this book–, which needs to be clarified: this is where we will introduce the notion of the Ontological Temptation.

My refusal to deal in objects and facts when articulating my stance (and thus, my rejection of non-deflated grounding and truthmaking) may sound obnoxious and unmotivated. But there is a rationale behind it. To see it, consider the following point. Cameron (2008) explicitly deployed truthmaking to deal with issues of ontological commitment. Suppose that we were to use this set-up for our purpose, viz. to frame Fine's (2001) distinction between what is really the case and what is merely the case, and thus to achieve a better understanding on issues of realism and anti-realism. This would have the consequence that all questions of realism somehow equate to issues of ontological commitment. Why is that? If, as I submitted, it is really the case that p if and only if "p" is true and is made true by the fact that p, then the fact that p is a truthmaker, and so, by Cameron's own lights, it is part of the "fundamental reality" from which we draw our ontological commitments. This holds true, mind you, even if we avoid Cameron's specific approach to ontological commitment in favour of something more traditional: if an issue of realism concerns itself with the issue of, say, "whether it is really the case that p", which, according to truthmaker theory, means that "p" is true and made true by p; then, for any non-deflated theory of truthmaking, from the truth of ““p” is made true by the fact that p”, we can existentially generalize to “there is the fact that p”; for Quinean metaontology, that would involve an ontological commitment to the fact that p. In conclusion, for someone who wants to treat Fine’s (2001) distinction through the lenses of non-deflated truthmaking, any question of realism ultimately amounts to an ontological matter – a debate about whether certain facts exist, and can function as truthmakers.

This is an awkward conclusion to reach. For it is only by way of logical torture that the moral anti-realist will concede that “killing babies for fun is wrong” involves unwanted ontological commitments. Maybe, as Fine (2001) suggested, from p we can safely infer that there is a fact that p; from that, we can infer that moral anti-realism, understood through the existential claim that there are moral facts, is false. But what if a moral anti-realist does not believe in facts? What if a moral realist wants to resist the inference from p to the fact that p? Fine’s Moorean “argument” against moral anti-realism will not go through, for the very simple reason that, on this understanding of the position, moral anti-realism should not be understood as the claim that there are no moral facts. Similarly, perhaps we can quantify at the second-order from “killing babies for fun is wrong” in order to put pressure on the moral anti-realist who claims that there are no moral properties. But then again, what if the moral anti-realist is a nominalist about properties, and does not wish to quantify at the second-order?

A moral realist need not embrace (moral) facts or properties, and conversely, the moral anti-realist need not reject them: their problem with “killing babies for fun is wrong” is not that the sentence is ontologically committing to some problematic entity; the problem is that, for lack of a better way to put it, it is characterizing reality along moral lines. The kind of misalignment – perceived by the moral anti-realist – between appearance and reality generated by truth of “killing babies for fun is wrong”, need not necessarily be expressed as a conflict between different ontological commitments. All in all, the issue of the reality of morality should not be understood as an issue of ontological commitment, albeit, perhaps, still a genuinely metaphysical issue.

Emboldened by the aforementioned “re-orientation” in metaphysics, one may want to shift the focus from ontological commitment to fundamentality; and thus claim that the issue of realism for morality should be understood as an issue about the fundamentality of moral facts such as the fact that killing babies for fun is wrong. But this is still unsatisfactory: if a moral realist does not believe in any substantive notion of fact, then surely they will not believe that moral facts can be fundamental.

Here is the heart of the problem: someone who insists on framing the issue of the reality of morality in terms of the existence or fundamentality of certain entities – be they facts, properties, or what have you – is indulging in an Ontological Temptation. Understanding this Ontological Temptation is crucial for the purpose of this book, and my discussion of Dispositional Reality in later chapters crucially relies on it. Roughly put: to indulge in the Ontological Temptation is to perceive an issue concerning the reality of certain aspects of the world (from numberhood to morality, from modality to lawhood and dispositionality, up to purely logical aspects such as negation or quantification) as an issue about the metaphysical status of certain entities – be it their existence, fundamentality, or what have you. The background idea behind the Ontological Temptation, in its broadest possible formulation, is that to talk about reality is to talk about certain entities that mind-independent reality is parsed into; ultimately, behind the Ontological Temptation is the "common tendency within reflective thought to be influenced, and even gripped, by a conception of the world as intrinsically 'divided' into discrete bodies" (Laycock, 2006, p. 3). Maybe the ensuing discussion is then framed as a discussion about which of these entities exist, or how they are arranged in the metaphysical scaffolding of the universe. But still – the discussion is about them.

Is this the correct way to pursue metaphysical discussions about reality? Is the Ontological Temptation harmless? Later in this chapter, in Sect. 2.6, I will deal with a strictly related topic, viz. the value of ontology vis-à-vis ideology in metaphysics; but lofty considerations aside, the case of morality already suggests a broad motivation for not giving up into the temptation: for not all questions concerning reality need be expressed through questions concerning the metaphysical status of certain entities. If my understanding of the issue of moral realism is correct, then the crux of the debate between the realist and the anti-realist will be about whether sentences with moral predicates can be perspicuous, viz. they reveal a moral aspect of reality's structure. Whether reality involves "moral entities" like moral facts or moral properties, is another matter entirely, one on which we should *ceteris paribus* be agnostic about. The same goes for other domains of discourse: we may want to discuss the perspicuity of modal talk – whether reality is inherently modal – without having to settle whether reality is inhabited by "modal entities" of any kind. Agnosticism seems proper here, at least for the time being. Debates like the one concerning moral realism (and, as we will see at length in this book, realism about dispositions) show that at least in some cases issues of realism can be separated from discussions about the metaphysical status of entities; if we can find a way to frame such issues without requiring to indulge in the Ontological Temptation, we will be able to avoid overextending on our metaphysical enquiry.[15] Here is another way to put it: someone indulging in the Ontological Temptation may commit what we may call an Ontological Fallacy, viz. failing to admit the reality of something because one cannot find, amongst the intuitive and familiar furniture of the world, the right entity for the job. We will see the Ontological Fallacy at work many times throughout this book.

[15] In other cases, admittedly, the Ontological Temptation appears to be less dangerous; e.g. it does not seem that big of a leap to go from a discussion about whether it is really the case that there is a tree in Lorenzo's backyard, to a discussion about the existence or fundamentality of a tree *qua* entity. Intuitively, trees are more likely inhabitants of reality than mysterious moral and modal entities. In my opinion, however, we shouldn't rely too much on this intuition; the possibility of framing certain realism debates without "going ontological" should be kept wide open in all cases. At any rate, the focus on this book is on the discussion concerning the reality of dispositions, which, just like moral realism, is one in which the Ontological Temptation is especially harmful; in the beginning of Chap. 4, when introducing dispositions, I will argue more at length in favour of this analogy.

I am not the only one to have voiced similar concerns. On this matter, the reader will perhaps forgive a brief detour: I was always taken aback by Lewis' decision to call his position about possible worlds "modal realism", an expression which has been more or less accepted by the ensuing literature. On the face of it, there is another position more deserving of that name: the position according to which modality is a real or objective matter. At best, Lewis' use of the expression is confusing. At worst, it may lead the inexperienced reader into a fallacy: that of believing that modal talk can only be taken in a realist spirit by paying an ontological price; or, more to the point, can only reveal metaphysical structure by revealing some component entity of reality. But just as one can be a moral realist without believing in special moral entities, one can be a realist about modality without believing in special modal entities – in short, one can be a realist about modality without being a modal realist. Relatedly, many objected to the apparent conflation between a certain domain of discourse being objective, and that domain of discourse being about certain special entities governing truth-conditions, e.g., Mondadori and Morton (1976, p. 5), and Yablo (2000, p. 9). Yablo wanted to separate ontology, viz. the study of which entities populate reality, from what he called "alethiology", viz. the study of what is the case – what I called "factuality"; this is perhaps closer to what I mean, with the proviso that "alethiology" is concerned with truths, whereas I am interested in something more: perspicuity. Yablo submitted that true sentences can reveal what is the case without revealing an ontology. I am making the stronger claim that, if the Ontological Temptation is to be resisted, (some) true sentence reveal metaphysical structure without hereby revealing an ontology.

But how exactly that is going to pan out? How can we understand the notion of structure and perspicuity, all the while resisting the Ontological Temptation? This is what we now (finally) turn to.

2.5 What Is Metaphysical Structure? And Perspicuity?

Here is a brief summary. Issues of realism should be understood as discussing what is really the case, as opposed to what is merely the case. In turn, this distinction should be understood as the claim that some sentences are perspicuous, and not mere truths. Realism and anti-realism about certain matters are thus to be reframed as matters of perspicuity, and, ultimately, structure.

The notion of metaphysical structure is therefore central here. Reality is not an amorphous lump, but it displays some kind of internal complexity which can be revealed by certain fragments of language. But what does that mean, exactly? What is structure? And what is it for a perspicuous sentence reveal it?

These are issues that we cannot postpone anymore. Two desiderata will guide our enquiry. The first one is that any suitable account of structure will avoid indulging in the Ontological Temptation. The second one is a desideratum of compositionality.

To understand it, consider by contrast an account of perspicuity which exhausts itself at the sentential level; perhaps, inspired by Fine's (2005a, p. 268) reality operator, for some sentential operator α, "p" is perspicuous iff α:p; or, mentioning the sentence rather than using it, for some condition $\phi(x)$, "p" is perspicuous iff $\phi(p)$. The problem here is that something important is lost when perspicuity is exhaustively expressed at the sentential level. By simply saying that α:p, or that $\phi(p)$, we are told nothing about the portion of metaphysical structure that p, as a perspicuous sentence, is supposed to reveal. That information should be conveyed by the sub-sentential components of p; in the sense that, in a perspicuous sentence, information about worldly complexity is conveyed through sub-sentential complexity of the sentence itself. E.g., take a sentence such as "electrons have electric charge -1e", and furthermore suppose that it is not only true, but metaphysically perspicuous: to say that electrons have electric charge -1e reveals something about the make-up of reality, about how things really are: what could that mean? Its perspicuity may have something to do with electrons, or electric charge; there's no single way to articulate this thought (we will shortly see several) – perhaps it has to do with the fact that reality includes such things as electrons, or the fact that electric charge is a genuine property, instead of a spurious one. It could be any of such things; thus, it would then seem that the sentence "electrons have electric charge -1e" is metaphysically perspicuous because its components (on a given interpretation), or perhaps the way they are assembled in the sentence, pick up some component or aspect of reality, which is what we have called metaphysical structure. We might say that perspicuity, like truth, is compositional, which is what ensures its connection with the metaphysical structure that a perspicuous sentence is supposed to reveal: only in a perspicuous sentence, metaphysical structure can be observed by studying the sentence itself, its components and perhaps their arrangement.[16]

With these desiderata in mind (avoid the Ontological Temptation, and compositionality), how should we understand the notion of metaphysical structure and perspicuity? In philosophical literature, there are two answers to these questions, and both can be associated to the work of two philosophers: the first is the approach to structure and perspicuity inspired by Logical Atomists, viz. with their own differences, by Ludwig Wittgenstein and Bertrand Russell; the second is a more recent approach developed by David Lewis and Ted Sider; I will call these two approaches the Wittgenstein-Russell Way to Structure and Perspicuity (for brevity, WR), and the Lewis-Sider Way to Structure and Perspicuity (LS).[17]

[16] On a similar note, Sider (2011, p. 148) argues against Fine's sentential approach by noticing the existence of what we may call "patterns of perspicuity", which may only be explained by appealing to the compositional nature of perspicuity.

[17] It may be worth stating that I am not attributing WR to Wittgenstein and Russell or LS to Lewis and Sider; they are reconstructions based on relevant positions they have submitted in their work. In other words: this is not an exegetical exercise.

2.5.1 The Wittgenstein-Russell Way to Structure and Perspicuity

Talk of metaphysical structure and perspicuity is not unprecedented in contemporary philosophy, if one knows where to look. Meet WR, the approach to structure and perspicuity inspired by Logical Atomism, as in Russell's *The Philosophy of Logical Atomism* (1918), and Wittgenstein's *Tractatus Logico-Philosophicus* (1922).

Logical Atomism is a multifaceted doctrine with several moving parts in semantics, epistemology, and metaphysics; there is no hope, and no need, to articulate it here in its entirety. For our purposes, we can concentrate on a couple of crucial points, mainly extrapolated from Russell's version of the position. First and foremost, according to Russell, atomic true sentences in a formalized language correspond to facts (Russell spoke of propositions as opposed to sentences; this distinction is not terribly important). WR is a way to cash out the idea that true atomic sentences describe reality by "carving up a piece of it" in the form of a fact. But this is not all: a true atomic sentence characterizes this piece of reality, this fact, by highlighting some of its internal complexity: facts have component entities, so-and-so arranged. According to WR, the world is ultimately a world of facts, which is what we talk about when we speak truly; furthermore, the world has a structure in the sense that these facts display some degree of internal structure. What is metaphysical perspicuity, then? According to WR, an atomic sentence is metaphysically perspicuous, or as Russell (1918, p. 25) would have called it, a part of a "logically perfect language", if and only if its syntactic complexity matches that of the corresponding fact: only to any component of the sentence, corresponds a component of the fact itself, and some structure is preserved between the former and the latter. In short, perspicuity amounts to isomorphism, and objective world-complexity can be revealed by observing the syntactic structure of the sentence itself (Cortens & O'Leary-Hawthorne, 1995, p. 155). Only with a logically perfect language we can perform what Russell had dubbed "logical analysis", and thus discover the "logical atoms", the ultimate building blocks of reality, as in Bostock (2012, p. 252).[18]

We must be careful, however. There are certain elements of Logical Atomism that are ill-suited for our purposes, and will need to be corrected (again, to which extent these corrections can be traced back to Logical Atomism proper is beyond the question – many of them, I suspect, cannot). First and foremost, we need perspicuity to be sufficient but not necessary for truth; yet there are reasons to think that the

[18] There are some differences between Russell's and Wittgenstein's brands of Logical Atomism. E. g., according to Wittgenstein the world, as a world of facts, displays structure to the extent that such facts are not simple but consist in configurations of objects (so-called "Tractarian objects"); for our current purposes, it may be worthwhile to point out that in Wittgenstein's logically perfect language there are no predicates: an atomic sentence is a concatenation of names (*Tractatus* 4.22). This difference bears some crucial consequences in the background metaphysics. E.g, it is unclear whether amongst the logical atoms, viz. the Tractarian objects, we will find entities corresponding to what we may call properties; and so, there is both a "nominalist" and a "non-nominalist" interpretation of Wittgenstein's Logical Atomism, as in Newman (2002, p. 62).

isomorphism sketched above fails in that regards. After all, we can turn a true (and perspicuous) sentence into a false one simply by switching a sub-sentential expression for another one in the same syntactic category; but if the former sentence is isomorphic to a correspondent fact, the latter one also is – despite being false. Equivalence classes of logically isomorphic sentences will comprise both true and false sentences, so the isomorphism in question cannot be used to define a notion of perspicuity that entails truth. Ultimately, this notion of perspicuity-as-isomorphism is a matter of syntactic structure, which was understood by Logical Atomism as a prerequisite for meaningfulness: every sentence isomorphic to a true sentence may not necessarily be true, but it is truth-apt, in the sense that there is a way the world could be for it to be true; it may in turn be isomorphic to a possible but non-obtaining state of affairs. There is a sense in which Logical Atomism conceived false sentences as revealing a certain kind of metaphysical structure, albeit not exactly as a portion of factual reality: it is revealing a portion in a space of possibilities. In conclusion, if perspicuity-as-isomorphism is simply a way for a sentence to match the complexity of some portion of reality, then it is safe to assume that Logical Atomists rejected the Factivity of Perspicuity.

This is clearly not the framework we are operating in. But I believe that we can salvage a non-trivial and philosophically interesting version of WR: according to it, the structure-preserving function that sanctions perspicuity is not just any function, but the intended interpretative function for the language in which the sentence is formulated (either formal or natural): when I wonder whether "Romeo loves Juliet" is perspicuous in this sense, I am wondering whether the entities corresponding to "Romeo" and "Juliet" stand in the relation expressed by the predicate "loves": if they do not, then the sentence is non-perspicuous, even thought, strictly speaking, the sentence may very well be isomorphic to a perspicuous true sentence. We might understand it as a correspondence-as-isomorphism theory, not for truth, but for perspicuity.[19] There is some intuitive pull behind this idea: whether a sentence accurately describes reality's structure is not a matter of naked logical form; it is the *interpreted* sentence to be perspicuous or not. And the sentence is perspicuous if the entities it is referring to are arranged in the way it describes.

This completes our elevator pitch for WR. It is based on well-known positions in the philosophical landscape; when Rayo was presenting Metaphysicalism^S, by his own admission he was presenting something in the vicinity of WR (Rayo, 2014, p. 512). The position has its virtues: most importantly, it easily satisfies the compositionality desideratum for perspicuity: whether a sentence counts as metaphysically perspicuous depends on whether its sub-sentential components satisfy certain conditions (viz., that they correspond to entities according to a certain structure-preserving transformation).

[19] What kind of theory of truth ought we to embrace then, that allows for a true sentence to be non-perspicuous? I will come back to this in the next chapter, where I will discuss more extensively the status of mere truths.

As they say however, the devil is in the details, and closer inspection will reveal WR to be a profoundly problematic position. The biggest problem for WR arises with the observation that the position comes with an obvious limitation: the notion of metaphysical perspicuity is only characterized relatively to atomic sentences. WR is tailormade for atomic sentences of an ideally formalized language, like $P(t_1...t_n)$; under a given interpretation, $t_1...t_n$ refer to a plurality of individuals, while predicate P expressed a certain property or relation, such that $P(t_1...t_n)$ is perspicuous if and only if those objects instantiate the property, or stand in the relation; the logical complexity of $P(t_1...t_n)$ then reveals some metaphysical structure – the objects bearing that property or standing in that relation. This approach does not need to be confined to atomic sentences in artificial languages; for example, we may take the sentence "Spain is west of France" as metaphysically perspicuous, to the extent in which "Spain" corresponds to an object, "France" to another, "is west of" corresponds to a relation, and the structure of predication in "Spain is west of France" is preserved as a structure of instantiation in the fact that Spain is west of France. WR primarily functions by exploiting the predication/instantiation isomorphism. But more complicated strings must to be considered, in which other kinds of syntactic components appear: quantified sentences, molecular sentences, modal sentences, etc.... How are we to cash out their metaphysical perspicuity? Logical Atomists notoriously had this problem concerning the truth rather than perspicuity of their sentences. Now the question is: given that, for our purposes, we must rule out a material equivalence between truth and perspicuity, how can we generalize WR to obtain a treatment of metaphysical perspicuity for *all* sentences? There are, as far as I can see, two ways of doing that, both of them severely problematic.

Firstly, we may model perspicuity in a way not dissimilar to truth; e.g. by making the perspicuity of a non-atomic sentence entirely a function of the perspicuity of its sentential components. In short, we might provide "perspicuity-tables". This iterative conception of metaphysical perspicuity renounces the background idea that a perspicuous sentence carves a correspondent piece of reality, in the form of a fact – this only holds true for atomic sentences. But it still allows us to say that a non-atomic perspicuous sentence reveals some structure about the world, even if all that reveals is what is already revealed by its components.

The biggest problem for this first option, it seems to me, is that while the semantic behaviour of logical operators is perfectly understood, the behaviour of such expressions vis-à-vis metaphysical perspicuity is not so clear. Therefore, the mileage on this approach may vary. Some cases are clearer than others; e.g., it seems harmless to suppose that perspicuity is closed under conjunction, and to claim that if both "p" and "q" perspicuously describe reality, then "$p \wedge q$" also does; whatever portion of metaphysical structure (viz., Russellian fact) is revealed by "$p \wedge q$" is also revealed by "p" and "q" alone. And so, we might want to say that "$p \wedge q$" is perspicuous if and only if both "p" and "q" are perspicuous.

But in other cases we lose grip on the notion of perspicuity that we are trying to cash out. Consider the case of disjunction; pursuing the analogy used so far, we might want to say that "$p \vee q$" is perspicuous if and only if either "p" is perspicuous, or "q" is; but that means that, given that "p" is perspicuous, "$p \vee q$" is perspicuous no matter how non-perspicuous "q" is. But is that true? Any decision on this matter seems arbitrary and poorly motivated.

An even more problematic case here would be negation. One might simply understand the perspicuity-clause of negation as stating: if p is perspicuous, ~p is not; and if p is not perspicuous, then ~p is. The first part of the clause is fine: if p is perspicuous, then it is true (by the Factivity of Perspicuity); and if p is true, then ~p is false; if ~p is false, then ~p is not perspicuous (by the Factivity of Perspicuity again). The problem is the second part of the clause: if p is not perspicuous, then ~p is. Surely this is not right. A sentence p may a non-perspicuous truth, but its negation ~p cannot itself be perspicuous, given that it is false. Now suppose that we were to amend, *ad hoc*, the perspicuity-clause of negation to state: if p is perspicuous, ~p is not (same as before); if p is not perspicuous (and true), then ~p is not perspicuous; if p is not perspicuous (and false), then ~p is perspicuous. The first two parts of the clause are correct. First: if p is perspicuous, ~p is not (as seen before). Second: if p is not perspicuous (and true), then ~p is not perspicuous. This is also seemingly unassailable: if p is not perspicuous, but true, then ~p will be false, and thus not perspicuous (by the Factivity of Perspicuity). The problematic part of the clause is the third: if p is not perspicuous (and false), then ~p is perspicuous. And so, according to this part of the clause, all it takes to generate perspicuity is to negate a falsity. E.g., suppose that for the moral anti-realism "killing babies for fun is wrong" is true, but not perspicuous. The sentence "killing babies for fun is right" is false; which means, according to the perspicuity-clause of negation just stated, that "killing babies for fun is not right" is not simply true, but metaphysically perspicuous. Keep in mind that the problem with the perspicuity for "killing babies for fun is not right" cannot consist in a failure of isomorphism; according to this option, perspicuity only requires isomorphism for atomic sentences, and "killing babies for fun is not right" is not an atomic sentence. But still one may have the feeling that to provide a metaphysically privileged description of reality, it is not merely sufficient to negate a falsity – which may cast some doubt on the very idea that perspicuity is entirely a function of the semantic action of negation.

To compound these difficulties, there is a further problem which this option inherits from the correspondent theory of truth, viz. that some non-atomic formulas do not have sentential components; which means that their perspicuity cannot be a function of the perspicuity of elements of its syntax. This is notably the case again of quantified sentences, which are very common amongst the premise of Moorean Argument that we have discussed so far (viz. "there is a prime number between 3 and 7", "there is a tree in Lorenzo's backyard", etc.. . .).[20] So let us move to the second option.

---

[20] It was in fact the basic insight behind Tarski's "The Concept of Truth in Formalized Languages" that it is impossible to define truth for quantified sentences in the same way as conjunctions or disjunctions, since the former, unlike the latter, do not contain any atomic sentences as components. That said, there have been efforts to explain the truth conditions of quantified sentences in terms of non-quantified sentences –e.g. Halbasch (1971).

According to this second option, we might generalize perspicuity in WR by striking away the restriction to atomic sentences in the original formulation. In short, we can bite the bullet and understand perspicuity as isomorphism for all sentences, not just the atomic ones. This means taking logical operators, quantifiers, and perhaps modal operators too, as sentential or sub-sentential components of a perspicuous sentence which correspond to a component of a fact.

There is a *prima facie* resistance to this radical idea, for it dictates that facts are also composed by special entities corresponding to expressions such as "not", "everything", or "necessarily" – which may strike one as *ad hoc* entities. What is more, their relation with the more mundane components of Russellian facts is also poorly understoood. E.g., Russell (2018) discusses what he calls "anti-quantificationalism", viz. the position according to which quantification-talk is never metaphysically perspicuous; when cashed out in terms of this variant of WR, the position claims that the facts corresponding to quantificational sentences do not have constituent entities corresponding to the quantifiers. But who would ever accept such a view? Any view of perspicuity that claims that sentences may reveal genuine structure by revealing the presence of dummy "quantification-entities" or "negation-entities" is gravely problematic.[21] The issue I have with this interpretation of WR is not, mind you, that such entities are mysterious, or strange. The problem is that it exacerbates a problem which is intrinsically built in WR's definition of perspicuity as isomorphism, viz. that it indulges too much in the Ontological Temptation, by settling issues of perspicuity (and thus, ultimately, issues of realism) by discussing whether (some, or all) sub-sentential components of a sentence correspond to out-there worldly entities. For the supporter of WR, a false dichotomy presents itself: either such entities are accepted in one's ontology, or the relevant issues of realism are settled in the negative. Consider the reality of logical complexity. Contrary to this radical interpretation of WR, Russell (1918, p. 39) and Wittgenstein (*Tractatus*, 4.0312) were adamant on the idea that there is no isomorphism for molecular sentences, and thus logical operators do not correspond to any out-there worldly entity.[22] This has occasionally been taken to entail that logical complexity belongs to the structure of language and/or thought; it is not a feature of the world. This would be a glaring instance of what I have called an Ontological

[21] Here is an exception, the reader might add: entities corresponding to modal operators. Just as truth-clauses for $\diamond$ and $\square$ involve possible worlds, their perspicuity-clauses might too. Possible worlds are not *ad hoc*, and can earn their keep, perhaps through a Lewisian cost-benefit analysis; furthermore, their relation with other components of the fact is, to some extent, well understood (e.g., as a mereological relation). Unfortunately, WR does not allow for the entities corresponding to $\diamond$ and $\square$ to be possible worlds, at least not as they are standardly understood. The isomorphism required for perspicuity dictates that both $\diamond$ and $\square$ are associated to a single entity each, and not to a domain over which to quantify over. Thus, if such things are possible worlds, there are only two of them: the "possibility-world" and the "necessity-world". This is not what possible worlds are.

[22] To be clear, Russell was famously hesitant on this matter: before his "Wittgensteinian phase", while presenting and defending his multiple relation theory of judgment, his stance on "logical objects" was noticeably different –e.g., in the 1913 manuscript "Theory of Knowledge", posthumously published in Russell (1984)–, although perhaps not completely clear (Griffin, 1985).

Fallacy: the idea that reality displays any degree of logical structure is discarded because the right ontology cannot be individuated. This second radical interpretation of WR may overturn this conclusion only by paying an ontological price: viz. accepting that such entities really exist, and correspond to logical operators in a molecular sentence. Unfortunately, both opinions on the reality of logical complexity are reached under the influence of the Ontological Temptation.

To be clear, the idea behind WR is not without its merits. Perspicuity as isomorphism functions well when we consider the contribution of certain sub-sentential expressions, but not others. More specifically, it seems to work well for individual constants, or, in natural languages, referential terms. Suppose that the sentence "Joe Biden is the US President" is metaphysically perspicuous. It is intuitive to think that the sentence reveals some portion of metaphysical structure because, amongst other things, "Joe Biden" corresponds to an entity inhabiting it. Here the correspondence between these components of natural language and the world has a name: reference. But the same thought is less intuitive when other syntactic categories are involved: it is far less intuitive that "Joe Biden is the US President" is perspicuous because (amongst other things) "is the US President" corresponds to an entity – presumably, the property being the US President. Perhaps "is the US President" corresponds to a set, and the isomorphism is between predication on the one hand, and set-membership on the other. Perhaps some other story can be concocted. By the point in which we arrive to sentential connectives, modal operators, and quantifiers, the analogy has completely broken down: to treat the role of those expressions like the role of referential terms is like to fit a square peg into a round hole. Furthermore, the starting point may be questioned as well: it may be intuitive to think that "Joe Biden is the US President" is metaphysically perspicuous because "Joe Biden" corresponds to a discrete entity inhabiting reality, but that doesn't mean it's true: after all, why couldn't an entity-free metaphysics perspicuously speak through individual constants? (I will not settle this matter here).

In short: for the supporter of WR, so understood, the Ontological Temptation manifests as a focus on a "quasi-referential" notion of correspondence when settling matters of perspicuity. Let's see if we can do better.

2.5.2 *The Lewis-Sider Way to Structure and Perspicuity*

LS displays the virtues of WR without its drawbacks; mostly, it preserves the compositionality of perspicuity without indulging in the Ontological Temptation. As such, it is a much preferable approach to structure and perspicuity, and one worth employing.

One may start to get a grip on LS by examining Lewis' notion of a natural property. According to Lewis (1983), properties as classes of *possibilia* adequately provide semantic values to predicates, but they are unsuited for other jobs – such as providing reality with structure, e.g. by ensuring objective similarities within classes and objective dissimilarities across classes. The reason is that there's so many of them, at least one for each predicate with a well-defined extension: "[p]roperties carve reality at the joints – and everywhere else as well. If it's distinctions we want, too much structure is no better than none" (Lewis, 1983, p. 346); a special entity, such an Armstrongian universal, can be posited to fulfil that role and provide the world with structure. Only some properties-as-classes are natural, in the sense that only for some of them, all the objects in the class share a universal. E.g., assuming that there is a universal of green, but not one of grue, while both sentences might be true "the ball is green" and "the ball is grue", only one of them is describing reality's structure. This vaguely approximates a notion of perspicuity. Yet it has two drawbacks; first: our desideratum of compositionality suggests that all sub-sentential components play a role in establishing the perspicuity of a sentence, but Lewis' notion of naturalness only works for predicates; second: the account is still excessively ontological to the extent that it populates reality with universals; when it comes to predicates, Lewis' naturalness still loosely resembles WR. But surely perspicuity, and structure, are not exclusive purview of the friend of universals.

Luckily for us, Sider's (2011) extension of this approach is explicitly designed to solve both problems. In short, the account begins with the introduction of a "joint-carving" operator $\mathbb{S}$ which can attach to sub-sentential expressions of any syntactic category, in either a formal or a natural language – to obtain things like: $\mathbb{S}$(green), $\mathbb{S}$(grue), $\mathbb{S}(\forall)$, $\mathbb{S}(\Box)$, etc.... There's no straightforward way to express $\mathbb{S}$ in a natural language, so we might approximate that by using a predicate applied to linguistic items; e.g. we might express $\mathbb{S}$(green) by saying that the predicate "green" is joint-carving; however the reader should not take this notation ontologically seriously (Sider, 2011, p. 108–109). Furthermore, there is a lesson from the difficulties of WR that ought to be remembered here: $\mathbb{S}$ does not simply apply to syntactic components of a sentence, but *interpreted* ones. After all, without the semantic interface between language and reality, expressions in the scope of $\mathbb{S}$ are simply syntactic strings, and it makes no sense to wonder which syntactic string is more perspicuous than another. In the case of predicates, it is only with an intended interpretation in mind that we can detect a distinction in naturalness between "green" and "grue" (the case of other sub-sentential components is slightly different, of course; quantifiers and connectives have a fixed semantic contribution which is built into the theory, rather than being model-dependent; so be it).

With these clarifications in mind, what does it mean to say that $\mathbb{S}$(green), viz. that the predicate "green" – in an intended interpretation– is joint-carving? Let me start by illustrating what it does *not* mean: the biggest departure from both Lewis' notion of naturalness, and WR, is that $\mathbb{S}$(green) does not state, nor entail, that the predicate "green" corresponds to a property, in any sense of the expression; similarly, to say that $\mathbb{S}(\forall)$ is not to talk about an alleged "quantification-entity". My doubts concerning the Ontological Temptation are clearly shared by Sider (2011, p. 111) who opposes his account to what he calls "ontologism", viz. the claim according to which "fundamental metaphysical commitments [are] ontic", and according to which "the best metaphysics of fundamentality is entity-based" (*ibid.*, p. 112). The adoption of an approach to structure and perspicuity freed from the shackles of ontology can be liberating (e.g., according to LS, freed form the Ontological Temptation, logical complexity may very well be a feature of the world, if logical operators are joint-carving).

That said, what does it mean to say that "green" is joint-carving, or that quantifiers or the logical operators are joint-carving, and thus part of the structure of reality? Ultimately, what is structure? The simplest way to understand the kind of structure afforded by this approach, comes by considering the notion of objective similarity. If it is the case that $\mathbb{S}$(green), and if is indeed the case that, say, the ball is green and the tree is green, then there is an objective similarity between the ball and the tree; in other words, the repetition of the word "green" in "the ball is green" and "the tree is green" is underpinned by a genuine pattern of similarity in the world. In this sense, reality has structure. To the extent in which the metaphysician is interested in studying reality and the patterns it exhibits, the use of the predicate "green" is metaphysically revealing; whereas that is presumably not the case for the "grue" predicate.

That said, the idea of similarity between objects, such as the ball and the tree, is just an approximation of the real thing; the reason is that, again, $\mathbb{S}$ is not confined to predicates, which are said of objects. Sider (2011, p. 88–89) suggests moving from object-similarity to fact-similarity. Roughly put, take α as an expression of an arbitrary syntactic category, and take two true sentences p and q in which α both appears; if it is the case that $\mathbb{S}$(α), then there is a genuine similarity between the fact that p and the fact that q. The repetition of "α" in both sentences is underpinned by a pattern of similarity between the two occurrences; and its use is metaphysically revealing. Please note that Sider's notion of fact-similarity does not require one to reify facts either, as we can use some sentential operator to cash out a non-ontic notion of similarity: to say that there is a genuine similarity between the fact that p and the fact that q amounts to claim that: when p, it's like when q (Sider, 2011, p. 105).

Some examples are as follows. Suppose that $\mathbb{S}$(green) and ~$\mathbb{S}$(grue); on this account, this should not be taken to mean that there is a universal of green, nor one of grue – nor it should be taken to mean that the property of grue is less fundamental the property of green. And so, issues of realism about whether it is really the case that things are green or grue, do not translate to metaphysical dispute about the existence, nature, fundamentality, or naturalness of these properties. Rather, the account functions as follows: given arbitrary t_1 and t_2 such that they are green, and arbitrary t_3 and t_4 such that they are grue: it is the case that when t_1 is green, is like when t_2 is green; but it is not the case that when t_3 is grue is like when t_4 is grue, viz. assuming that $\mathbb{S}$(green) and ~$\mathbb{S}$(grue), there's an objective similarity between the former two facts, but not between the latter two, which can be picked up by the repetition of the joint-carving expression. Similarly for the logical operators or quantifiers; e.g. conjunction. Claiming that $\mathbb{S}$(and) will amount to claiming that, e.g.: when it is sunny and windy is like when it is cloudy and rainy. Most importantly, this does not state nor entail that there is an out-there conjunction-entity that makes up for the similarity. It only states that there are genuine objective similarities out there which can be picked up by the repetition of the joint-carving expression. And the same goes for the other sub-sentential expressions. This will also be the case, as I will submit, for dispositional ideology.[23]

Building from Lewis' and Sider's strategies, we can finally present an account of perspicuity and structure which satisfies our desiderata. According to LS a sentence is perspicuous if and only if it is true and formulated in joint-carving terms; because the 𝕊 operator attaches to expressions of any syntactic category, all subsentential components are supposed to play their part in ensuring the perspicuity of a sentence; furthermore, 𝕊 does not require that the expression corresponds to an entity to be perspicuous. In short, LS is suitably compositional and suitably non-ontic, as required. Incidentally, the compositionality of perspicuity allows for a nice feature: we can detect the source of a failure in perspicuity – given that the perspicuity of a sentence ultimately depends on features of its syntactical elements. E.g., for a moral anti-realist all descriptive sentences containing "is wrong" will be, at best, merely true, presumably because ~𝕊(is wrong), which is the ultimate source of their moral anti-realism, together, presumably, with ~𝕊(is right).

[23] Here is an objection against the notion of fact-similarity, and its role in the context of this book. My metaphysical background does not envision a heavyweight notion of fact to separate reality from appearances, and rather employs a deflated/abundant conception of fact: if it is the case/it is true that p, then there's a fact that p. Does that mean that these fact-similarities are exhausted at the level of appearances –how the world appears to us?

My answer would be: no. First, the realm of facts is not simply the realm of appearances as opposed to reality. Since perspicuity is a subset of truth, sometimes it appears to be the case that p, but it is also *really* the case that p –as I will discuss in the next chapter, not all appearances are *mere* appearances. In other words, the realm of facts is simply the realm of truths, or factuality. Therefore, "…is like when…" is not merely registering similarities in the "world of appearances" (perhaps at the linguistic, representational, or conceptual level) for the simple reason that the sentences flanking the operator are used, rather than mentioned. In describing fact-similarities, we are talking about the world, not any representation of it, perfect or imperfect as it might be. Fact-similarities are real –a genuine part of an objective mind-independent reality: one does not need to reify facts, nor dispositions, to get there.

I can offer an analogy, anticipating a notion that will come back multiple times throughout this book: one can think of fact-similarity as a close relative of generalized identity just-is operators. They are similar to the extent that they are both sentential connectives (as opposed to predicates) that do not require the reification of the "bits of reality" that they are saying are similar (in my case), or identical (in the case of generalized identity). When a supporter of generalized identity says that "for the number of dinosaurs to be zero *just is* for there to be no dinosaurs", they are not assuming the existence of any heavyweight fact (nor numbers, nor dinosaurs, for that matter) –after all, again, the sentences are used, rather than mentioned, to compose the overall sentence–, to express something objective about reality, in this case a "sameness in content". I am doing something very similar, relating to a "similarity in content" instead. Friends of generalized identity like to call their operator a "no difference" operator, and sometimes struggle to informally characterize it in an illuminating way: what they usually say is that it is a categorical extension of *objectual* identity. My fact-similarity operator is in a similar boat: it is categorical extension of *objectual* similarity, to the extent that does not require the "objectification" of two bits of reality that are said to be similar. It simply says that: when the world is like this, and when the world is like that, there is a similarity between these two ways the world is.

What about structure itself, then? According to WR, structure mostly amounts to the internal complexity of facts, while according to Schaffer (2009) it amounts to the layering imposed upon the world by a relation of grounding. But for LS metaphysical structure is not a thing,[24] and as I have explained, we should not presuppose that $\mathbb{S}$ is parsing reality in discrete units neither at the sentential nor at the subsentential level. The claim that reality has structure does not equate with the claim that it is composed by objects, properties, relations, facts or what have you. What is going on is rather different, in the sense that the perspicuity of a sentence informs us about the structure of reality in a slightly more elusive way. Suppose that "R(a)" is both true and perspicuous, and thus $\mathbb{S}$(R) and $\mathbb{S}$(a); given that $\mathbb{S}$(R), for any true sentence p^R in which "R" appears, when p^R is like when R(a); and similarly, given that $\mathbb{S}$(a), for any true sentence p^a in which "a" appears, when p^a is like when R(a). Knowing that "R (a)" is perspicuous thus amounts to knowing that the fact that R(a) is part of a complex pattern of similarity underpinning the use of both "R" and "a". This is, in a way, novel information which is not simply obtained by knowing that $\mathbb{S}$(R) and $\mathbb{S}$(a).[25] To help the reader see the point, allow me to briefly break my own rules and go ontological: suppose that true sentences do correspond to facts; the joint-carving nature of their subsentential components, so understood, entails that these facts can be arranged in similarity classes; each of those classes constitutes a joint of reality; assuming that $\mathbb{S}$(R), and that it is the case that R(a), R(b), R(c), etc.... then the fact that R(a), the fact that R(b), the fact that R(c)... etc., are all members of the joint corresponding to R, viz. the R-joint. And the same with $\mathbb{S}$(a). So understood, the fact that the true sentence "R(a)" is perspicuous (viz. it is composed by joint-carving expressions R and a) is telling us something important about reality, viz. that the R-joint and the a-joint have a non-empty intersection in the fact that R(a). This is the novel information I was talking about before. That said, I do not really wish to go ontological, and I do not wish to reify facts or joints. So this was but a vague gesture: the ontologically deflated notion of fact-similarity used so far still allows us to say that if "R(a)" is perspicuous, there are similarities occurring between the fact that R (a) and any other fact described with the use of the predicate R and the constant a.

Another peculiarity of this approach to perspicuity is emerging quite clearly: the compositionality of perspicuity is indifferent to logical form: only the joint-carvingness of the subsentential components matters, but not how they are arranged in a grammatical structure. Mind you, sentences entirely composed joint-carving expressions can fail to be perspicuous, if they are false (Factivity of Perspicuity); but if, say, $\mathbb{S}$(R) and $\mathbb{S}$(a), and if "R(a)" is true, then there is no further question as to whether R(a) is metaphysically perspicuous: it is. In my opinion, this is a feature rather than a bug. After all, $\mathbb{S}$ functions by ignoring grammatical categorizations, so it isn't peculiar that the emerging feature of perspicuity is also grammatically-

---

[24] Sider (2011, p. 6, fn 5), but also see the exchange in Contessa (2013a, b) and Sider (2013).

[25] Similarity is standardly symmetric but non-transitive; this precludes the uncomfortable conclusion that, given the perspicuity of "R(a)", for every p^R and every p^a, when p^R is like when p^a.

insensitive. But there is a deeper reason for this move: I do not believe that we should read metaphysical structure off syntactic structure. The litmus test here lies in the possibility, if not actual existence, of fundamentally alien languages, viz. languages that do not simply involve alien nouns and predicates, but display a completely different grammatical structure. There are languages without nouns, as in Rayo (2013, p. 13), or as in the famous story *Tlön, Uqbar, Orbis Tertius* (Borges, 1962); and then there are languages without predicates, like Sellars' (1962) Jumblese. The possibilities are endless. Yet there's little indication that sentences in these languages cannot reveal the same portions of metaphysical structure than more ordinary sentences. Whether or not there are individuals, or properties, or both, or neither, fundamentally different languages can still be understood as different ways to express the same structure. How exactly this "metaphysical equivalence" can be established and expressed, goes well beyond the scope of this book, but if we tie perspicuity to grammatical structure, the project is dead on arrival. So I stand by the claim that the compositionality of metaphysical perspicuity is indifferent to syntactic structure.

Finally, I would like to address a potential source of confusion concerning LS. To claim that, e.g. $\mathbb{S}$(green) is to make a claim about reality, and more precisely, a claim about a privileged correlation between a linguistic expression (in a given interpretation) and reality, viz. what we have been somewhat metaphorically describing as "carving at the joints". Can we say something more on this relation? Sider (2011) is occasionally cryptic on this topic; he insists that structure, and $\mathbb{S}$, are to be taken as primitives of the theory, although they can be characterized as performing certain theoretical roles, in this case, that of ensuring similarities. But *what* exactly is responsible for the similarity? Suppose that $\mathbb{S}$(green), and that, therefore, when the apple is green is like when the lime is green: whence the similarity? *Prima facie*, one may believe that "green" is joint-carving if and only if reality exhibits the suitable joints (whatever joints are) – which have a special correlation with the predicate "green", and which ensure the objective similarity between the two aforementioned facts. Of course, when formulating this alternative, we must be very careful not to crystallize these joints into entities, or this whole detour into LS to avoid the pitfalls of WR would have been in vain; but this temptation is, once again, poorly motivated: there may be plenty of ways to interface the predicate "green" with reality, in a way that does not entail correspondence with a worldly entity (to think otherwise is to indulge once again in the Ontological Temptation, viz. to think that joint-carvingness can only be real if there are entities which are "carved" by the joint-carving expression.) Similarly, there are many ways for $\mathbb{S}$(green) to ensure a worldly similarity that does not involve the re-occurrence of a greenness-entity (like the universalists think) between the two facts. For the moment, I would like the account to remain neutral on these matters.

Perhaps may follow Sider (2011) in his primitivist stipulation – according to which, when we claim that green belongs to the structure of reality, this amounts to the fact that $\mathbb{S}$(green), and, therefore, suitable similarities can be established between various facts.[26] On this reading of metaphysical structure, the only privileged relation with reality that "green" enjoys over "grue" (assuming that 𝕊(green) and ~𝕊(grue)) is that "green" draws a distinction where there is one, whereas "grue" does not; the repetition of the "green" predicate reveals a genuine pattern of similarity in objective reality, whereas the repetition of "grue" does not. For our current purposes, Sider's primitivism seems like a good place to stop. There is no better way to understand the idea that reality has a structure – that it is not an amorphous lump – than the idea that reality exhibits certain patterns of similarities, which certain linguistic expressions, in a given interpretation, are particularly apt to capture.

2.6 Spurious Realism, and the Value of Ideology

Before concluding this chapter, there is a point that requires addressing. I claimed multiple times that I am not a deflationist in metaphysics, and that I do in fact believe disputes in metaphysics, including the dispute on dispositions, to be genuine debates about a mind-independent reality; my acceptance of metaphysical structure, and my endorsement of a type (C) compatibilist stance on Moorean Arguments, is meant to walk the narrow path between error theory and deflationism and thus rescue metaphysics as a worthwhile endeavour.

A suspicious reader may wonder whether this is really earned: is my account of structure and perspicuity sufficient to warrant the conclusion that I am metaphysical realist – not simply in the sense that there is a mind-independent reality, but in the sense that metaphysics is an investigation about it? Or perhaps I am stretching the notion of realism too much here? We may call this the Spurious Realist objection. It seems to me that the operative difference between deflationists and non-deflationists in metaphysics is methodological, and I will discuss the issue of methodology in the next chapter. Yet there are components of the position presented so far which may warrant this charge already. Firstly, there's the Ontological Temptation, viz. the claim that we shouldn't straightforwardly equate any debate in metaphysics as a debate about the existence or nature of a special class of ontological furniture. Is this, in a very general sense, already a way of conceding too much to the metaphysical antirealist? In the context of this chapter, non-deflationism about metaphysics is understood as the claim that mind-independent reality is a certain way – it has a

[26] Sider clarifies that similarity-making is not the only use he has for structure; and that, in fact, the best argument in favour of structure is that it is a unifying posit across a wide variety of philosophical phenomena (Sider, 2011, p. 13). In what follows I will not rely on this argument for metaphysical structure. My adoption of the notion is vaguely instrumental, in the sense that it is motivated by the need to properly distinguish between reality and appearance, to provide a compatibilist strategy for Moorean Arguments –a strategy Sider (2014, p. 416) himself acknowledges– and, relatedly, as we will see in Chap. 4, to properly distinguish between metaphysically naïve and metaphysically loaded dispositional talk.

certain nature, in the broadest understanding of the term– to warrant a non-trivial investigation of reality through the means of metaphysics. But to my mind, this doesn't entail that reality has to be parsed in discrete units functioning as variable-values or truthmakers. This is not to straightforwardly say that it isn't, but that debates in metaphysics, about the reality and nature of this and that aspect of our experience, from morality to dispositions, ought not to necessarily take that form.

The possibility of a moral anti-realist rejecting the existence of heavyweight (moral) properties or facts is, I believe, hinting in the direction of this extended logical space; another hint is the possibility of a friend of dispositions who is also a nominalist about properties. This extended space of position is generated by the logical gap between thinking that metaphysical issues are genuine issues about the reality of certain aspects of the world, on the one hand, and thinking that such aspects are instantly reified into their own special class of objects, properties, or facts. To be sure, I concede that this logical gap is often, albeit not always, ignored (viz. my account is not strictly hermeneutical in character, but rather revolutionary). But it is still there: my adoption of this Sider-inspired non-ontic notion of structure is a way to cash out the idea that moral predication fails to "carve at the joints" in a less straightforward fashion. Now, a charge of Spurious Realism would stick if it were to argue, in a non-question-begging capacity, that my non-ontic notion of metaphysical structure is unsuitable to characterize this nature of reality. I do not exclude that this argument may exist, but I am not aware of any version of it.

Secondly, the charge of Spurious Realism may rear its head by considering the opposition between the ontology and ideology of at theory, in a standard Quinean sense. By warning the reader against the Ontological Temptation, I have claimed that not all metaphysical disputes ought to be understood as disputes about ontological commitment; after all, debates about the morality of good and evil (but also debates about the reality of dispositions) usually revolve around expressions that we can find in a predicative position. By refusing to reduce these disputes to disputes of ontological commitment, I am taking at face value the idea that they primarily concern the ideology, as opposed to the ontology of a theory. Doesn't that mean that we are dealing with ideas, concepts, or other representational artifacts, as opposed to objective reality? At least, this is how the ontology vs. ideology opposition has been understood in the Quinean tradition (whether Quine himself would agree, I will not say). This seems enough to kickstart the objection of Spurious Realism. e.g., by claiming that realism about morality ought to be understood as concerning the nature and features of moral predicates, I am already undermining its credentials as a genuinely realist position: in my framework, moral realism is only *spuriously* realist.

Someone who has followed me so far will probably be able to imagine my response. In short, I do not believe that metaphysics is exhausted by ontology and that, once ontological commitment is put on the side, there is nothing of importance for the metaphysician to discuss concerning the ideology of a theory. My theory of metaphysical structure and perspicuity is but the latest in a growing trend of contemporary philosophy that takes there to be a genuine metaphysical value in ideology choice. The problem, as I understand it, is that taking choices in ideology as merely dictated by epistemic or pragmatic reasons has a trivializing effect in metaphysics (Finocchiaro, 2021). Trade-offs between ontology and ideology are commonly understood, as per the tradition above, as trade-offs between worldly complexity and representational complexity; e.g. David Lewis is famous for his willingness to pay for a slimmer theory with the coin of a (quantitative) ontological inflation, famously resulting in a world as vast and flat as a desert. But something is not right here. First of all, these trade-offs are a little bit too easy to come by. Finocchiaro (2021) proposes several examples, but Rayo (2013, p. 13) submits the more general point that, with the right logical tools, and thus the right theory, object-talk is eminently dispensable. If ontological parsimony is the only metaphysical value attached to a theory (understood as a body of sentences), then the metaphysician always has at their disposal the most parsimonious theory there is: one describing an empty world. Perhaps the metaphysician will respond: maybe so, but this theory will probably not be the *epistemically best* theory, thus I do not feel compelled to buy into its ontological commitments. In other words, they are appealing to epistemic or borderline pragmatic reasons to not posit an empty world. But this is not a good spot for the metaphysician to be: after all, reality is what it is, independently of our conceptualizations or representational efforts. If a theory is more ontologically parsimonious, but more complicated, less pragmatically appealing or more difficult to teach, aesthetically unpleasing – so be it. Whoever promised us that reality was easy to describe? In other words, the problem is that in in reducing the trade-offs between ontology and ideology to trade-offs between worldly complexity and representational complexity, we are weighting off metaphysical virtues such as ontological parsimony against epistemic or pragmatic virtues. Sider (2011, p. 14) claims that this would be tantamount of trading "apples for oranges"; to me, the problem is particular stringent for the metaphysician who genuinely believes in metaphysical realism, viz. in the claim that there is a reality independent of our conceptualization or representational efforts: in that case, psychologized ideology would never be able to provide a counterpoint to the massive ontological deflations afforded by the dispensability of object-talk.

Again, I am a metaphysical realist. And I do not believe the world to be empty. So I must accept that ideology choice must have some kind of metaphysical value.[27] This is what this chapter is trying to achieve with talk of metaphysical perspicuity. Under this new light, we can see that traditional trade-offs between ontology and ideology do not achieve very much: not much is changing in one's characterization of reality by switching back and forth between "it's storming outside" and "there's a storm outside", even though (taking surface grammar at face value) only one of these sentences is ontologically committed to storms. In both cases we are describing a world with storms in it. It is unwarranted to believe that the idea of a reality with storms in it can exclusively be expressed through the apparatus of quantification: most often than not such an apparatus can be easily dispensed with, in a way that would trivialize the efforts of the metaphysician (which is why, incidentally, it was

[27] This is what Finocchiaro (2021) calls *fidelic virtues*.

important for the 𝕊 joint-carving operator to be insensitive to grammar). And to be clear, Rayo's point about the dispensability of object-talk shouldn't be taken to entail that there are no objects, but that the metaphysical value of sentences shouldn't be exhausted by the ontological commitments that we can extrapolate through quantification.

This is not the first time we stumble upon this phenomenon, either. Remember the trouble we encountered characterizing the argument against moral anti-realism starting with "killing babies for fun is wrong". There are various ways to fill in the gaps here, either by way of some fact-quantification principle (from "killing babies for fun is wrong" to "there is a fact that killing babies for fun is wrong") or second-order quantification (from "killing babies for fun is wrong" to "there is wrongness"); but people who reject all of the aforementioned quantification-based inferences, can still count themselves as anti-realist (which is why the Moorean Argument was ultimately not very convincing). In other words, the question of realism concerning morality is orthogonal to any dispute concerning ontological commitment.

Far from being spuriously realist, my theory of metaphysical structure and perspicuity as a feature of sentences, and ultimately sub-sentential linguistic expressions (including ideology) is an explicit effort to extend the metaphysical value of a sentence, or of a theory, far beyond the narrow confines of ontological commitment. To borrow a slogan: "ideology matters" (Sider, 2011, p. i).

2.7 Conclusions

Metaphysical debates of the realism/anti-realism variety can be understood as non-trivial debates about what is really the case as opposed to what is merely the case; this distinction, in turn, can be cashed out as the distinction between metaphysically perspicuous sentences and merely true sentences. So understood, metaphysics as an investigation of reality mostly concerns itself with metaphysical structure. But we must be careful when talking about structure. According to LS, reality having a structure mostly amounts to reality exhibiting patterns of objective similarity which are picked out by a privileged subclass of linguistic expressions. Perspicuous sentences need not "carve out" reality in objects, properties, or facts. More generally metaphysics, as an investigation of a structured world, is not necessarily the investigation of a world parsed in discrete units, an assumption which is just as common as it is poorly motivated. So one should not pursue structure by trying to match subsentential expressions in a perspicuous sentence with component entities of reality, as the Logical Atomists did: as we have seen, this matching-strategy starts to behave awkwardly as soon as the leave the warm embrace of non-quantified atomic sentences.

Two final clarifications are needed here. One: this is not to say that LS, and by extension, the rest of this book, are off-limits for those who wish to postulate objects, properties, relations, facts, etc.... The point is rather that their postulation is orthogonal to matters of realism; some examples that we have offered may make the point particularly clear: e.g., the reality of morality, modality, and logical complexity. In the case of dispositionality, we may think of matters of realism as orthogonal to the postulation of a dispositional ontology: if one has independent motivations to posit such entities, then they are more than free to do it. These are further details to the structure of reality that may be warranted if additional philosophical pull is present; but are not presupposed in this book.

Two: amongst other things, LS separates matters of realism from matters of ontological commitment; but this does not entail that LS is a way to get realisms “for free”. Metaphysics, as a study of structure, is still meant to be a non-trivial endeavor. Realism has to be earned. Being committed to some fragment of language being joint-carving, and thus some sentences being perspicuous is an important commitment about the make-up of reality, albeit not an ontological commitment. Thus, on the primitivist variant of LS that I have adopted, one makes an important commitment about reality when claiming that some fragments of language are joint-carving.

How exactly are realisms to be “earned”, is a methodological point that we will discuss in the next chapter, where I will explain how to articulate the distinction between reality and appearance, and how to go on about discovering which is which.

Chapter 3
Appearance

3.1 Introduction

So much for reality – what about appearance? Questions of realism, we have submitted, can be understood as questions concerning the metaphysical perspicuity, as opposed to mere truth, of sentences. A key component of this approach is that some sentences may be true without being perspicuous: mere truths, describing what is merely the case as opposed to what is really the case, outline what we may call a "realm of mere appearances". But do not worry: this mysterious realm is not a real place.

An account of appearances, and its relation with reality, will complete my metaphilosophical framework, offering some important insight on the discipline of metaphysics, and suggesting an inkling for a philosophical methodology that will guide us in the rest for the book.

3.2 Against a Non-factive Conception of Appearances

As we have seen, a good reason to deploy the notion of metaphysical structure and perspicuity is to safely navigate between the Scylla of deflationism and the Charybdis of error-theory, viz. to have a compatibilist way out of Moorean Arguments that does not compromise the substantiality of metaphysical debates, without renouncing the truth of so-called Moorean certainties.

There is a price to pay however, viz. the claim that some sentences may be true without being perspicuous. In Fine's terms, this means that, given what is the case, there is a further distinction between what is really the case, and what is merely the case. As Fine (2001, p. 8) himself recognizes, this is a potentially mysterious claim to make: for it seems to be partitioning the world surrounding us between a reality and a "realm of mere appearances" with a somehow diminished metaphysical status; and contrariwise, the facts of reality enjoy a special status of some kind.

L. Azzano, *Dispositional Reality*, Synthese Library 482,
https://doi.org/10.1007/978-3-031-52625-1_3

Someone playing devil's advocate could take advantage of this difficulty to reestablish option (B) as a viable way to deal with Moorean Arguments (namely, by rejecting ordinary premises such as "there is a prime number between 3 and 7"). Although option (B), as a kind of error-theory, is not without its problems, it may be an inviting proposition to the extent that it reconstructs the difference between reality and appearances, as the distinction between true and false sentences; this immediately neutralizes all metaphysical threats concerning any mysterious "realm of appearances". According to the supporter of option (B), if there is any genuine conflict between reality and appearances, then this ultimately amounts to the distinction between true and false sentences; the potentially problematic distinction between what is really the case and what is merely the case deflates to the distinction between what is the case and what is not the case (which is, if not perfectly understood, at least better understood). Following Lipman (2018), we may take this error-theoretic approach as endorsing a "non-factive conception of appearances", according to which, schematically:

A_1) it is really the case that ϕ iff ϕ
A_2) ~(if it appears to be the case that ϕ, then ϕ)

This contrasts my approach, which involves what we may call a "factive conception of appearances", according to which

A3) if it appears to be the case that ϕ, then ϕ
A4) ~(if ϕ, then it is really the case that ϕ)

(Please note that A_3, unlike A_1, is not a biconditional; whether for every ϕ that is the case, it appears to be the case that ϕ – in any sense of "it appears to be the case", is a far more controversial claim that I do not intend to settle here). According to my factive conception of appearance, the distinction between reality and appearance should not be understood as the distinction between truth and falsity (or as the distinction between factuality and non-factuality), but as the distinction between metaphysical perspicuity and mere truth.

It is now time to provide some additional motivation for my choice (A_3–A_4 instead of A_1–A_2); more specifically, I would like to offer some independent reason for taking the premises of Moorean Arguments as true, thus ultimately arguing against option (B) and a non-factive conception of appearance.

3.2.1 *Moorean Certainties and "Lifting-the-Veil"*

The premises of Moorean Arguments describe a densely populated world: numbers, trees, wrongdoings, and much more: e.g., ways ("there is an easy way to do it"), times ("there is a time for everything"), ideas ("there are many ideas to improve things"), and so forth. It is the world of our everyday life, which we encounter and navigate while abstaining from any further investigation of reality and its make-up.

This last claim is, by itself, quite controversial. To probe the depths of Loch Ness and discover that there is no monster after all is, in a broad but incontrovertible sense, an investigation of reality; but surely one would not add "there is a Loch Ness monster" to the inventory of Moorean certainties – despite the fact that, abstaining from such an investigation, some would be prone to believe it. So why couldn't this be the case for the many Moorean Arguments considered in the previous chapter? E.g., could not numbers and wrongdoings – for the anti-realist at least– be like the Loch Ness monster: viz. something whose reality we might be willing to endorse at first, but not at later stages of investigation? This is the starting point for a hypothetical supporter of option (B) endorsing (A_1–A_2). Such a philosopher endorses what I like to call a "lifting-the-veil" conception of metaphysics, according to which metaphysics, as a theoretical investigation about reality, has the goal to re-evaluate and perhaps even discard ordinary and pre-theoretical beliefs. This is a flattering conception of metaphysics, that makes the discipline looks revelatory and sometimes even transgressive in nature. It is also, in my opinion, severely flawed. But it is important to understand why.

An important clarification to make at this point is that there is an epistemic or perceptive reading of "appearance" and "it appears to be the case that..." locutions which might warrant a non-factive conception of appearance. After all, it might appear that a stick half-immersed in water is bent, in the sense that someone might have apparently perceived it as such, although that isn't the case. And something similar goes, I suspect, for the Loch Ness monster. This perceptive reading of appearance goes hand-in-hand with the lifting-the-veil conception of metaphysics, although how exactly the case of, say, morality, or numbers, or dispositions, can be modelled after the case of the apparently bent stick, remains to be seen. On this approach, to reject the non-factivity of appearance would require rejecting the claim that perceptive errors are genuine instances of perception (perhaps embracing some kind of disjunctivism).[1]

I, for one, am not particularly interested in solving this puzzle, because I reject this perceptive reading of appearances in its entirety. More to the point, this perceptive reading of appearances is not the one I am primarily interested in: whether disjunctivism in theory of perception is true, it doesn't seem to me that the mereological nihilist should consider the tree in Lorenzo's backyard as a kind of misperception; and similarly for the moral anti-realist with the wrongness of killing babies for fun. The issue with the kind of anti-realisms that we are considering here, including anti-realism about dispositions, is not inherently perceptive in nature. Anti-realists are not usually concerned with the reality of mirages or optical illusions, but with the reality of phenomena that are widely recognized to be factual and describable in true sentences, such as "gold is highly malleable", "there is a prime number between 3 and 7", "there is a tree in Lorenzo's backyard", and so forth. In other words, we shouldn't take this notion of appearance to have epistemic or

[1] There also is an epistemic reading of metaphysical perspicuity that is mostly unintended on the present account: a perspicuous sentence is one whose logical structure gives us epistemic access to the structure of reality; I suspect that there are shades of this epistemic reading of perspicuity in the Logical Atomist's notion of a "logically ideal language" and the corresponding propositions.

perceptive connotations: an appearance, in this sense, is merely a Moorean certainty, an ordinary belief which constitutes the starting point for metaphysical investigation. The world of appearance is the realm of trees, persons, emotions, good and evil, numbers, times and places: it is not the realm of hallucinations and mirages. This may be a little loose as a distinction, but I take it to have intuitive grasp enough to function. And to anticipate things a bit, my argument for the factivity of appearances is not epistemic, but metasemantic.

To proceed we must, first and foremost, understand what a Moorean certainty is. As we have seen in the previous chapter, there is no overly stringent definition: in general, a Moorean certainty is something "we ordinarily believe" (Fine, 2001, p. 2), or a piece of "everyday knowledge" (Lewis, 1996, p. 549). Lewis' use of "knowledge", as opposed to "belief" immediately suggests another feature of Moorean certainties: that they are unassailably true pieces of information, albeit very shallow ones; in a way, they are unassailable true exactly because of how shallow they are: and it is often stated that Moorean certainties are more certain than the premise of any philosophical argument to the contrary (Lewis, 1996, p. 549; Schaffer, 2009, p. 357). These kinds of pronouncements are usually meant to re-orient our approach to Moorean Arguments in a certain direction, viz. away from strategy (B) and towards either (A) or (C). But is the truth of the premises of Moorean Arguments truly unassailable?

Reference to Moore's arguments against skepticism is probably misplaced here, since the term "Moorean certainty" as it exists in the literature is more likely a misnomer. Moore's argument for the existence of an outside world is not meant to be a shallow or trivializing argument (it is an argument against skepticism, after all), and the relevant premises are not meant to be certain simply by virtue of their being common sense or obvious, in whatever sense of the world. The certainty Moore was talking about is a more complicated affair, and has rather to do with the fact that such premises couldn't be truthfully denied, or could only be denied at the price of a logical or practical contradiction – or a combination thereof (Skirry, 2003).

The notion of a Moorean certainty in contemporary literature is perhaps more concerned with the methodologically disadvantaged status of philosophy vis-à-vis neighbouring disciplines from which the relevant sentences draw from. It wouldn't really make sense to reconsider arithmetical claims such as "there is a prime number between 3 and 7" on metaphysical grounds; famous, in this regard, is the pronouncement in Lewis (1991, p. 58–59). Yet, as intuitively appealing as this line of reasoning might be, I won't rely too much on it: for it strikes me as an excessively top-down tactic, too concerned with excessively vague matters about what philosophy is or isn't.

On the contrary, I will argue that to take these so-called "Moorean certainties" to be false would amount to bad linguistic practice, if not outright semantic error: in other words, I endorse a kind of metasemantic argument against A_1–A_2, and thus in favour of A_3–A_4. We may begin by observing that this lifting-the-veil conception, the idea that theoretical endeavours are in the business of refuting ordinarily held beliefs, is not exclusive to metaphysics. Scientists are fond of it too, especially in their disseminative efforts. Perhaps the most well-known example here consists in the debate between Eddington and Stebbing concerning the solidity of wood. In his *Nature of the Physical World*, Eddington famously wrote:

> I am standing on the threshold about to enter a room. It is a complicated business. In the first place I must shove against an atmosphere pressing with a force of fourteen pounds on every square inch of my body. I must make sure of landing on a plank traveling at twenty miles a second round the sun – a fraction of a second too early or too late, the plank would be miles away. I must do this whilst hanging from a round planet head outward into space, and with a wind of aether blowing at no one knows how many miles a second through every interstice of my body. *The plank has no solidity of substance.* To step on it is like stepping on a swarm of flies. (Eddington, 1928, p. 342, emphasis mine)

The passage is supposed to illustrate how the world is different when seen from a scientist's perspective; and Eddington's down-to-Earth analogies between a supposedly solid objects and solar systems, in which minute particles are separated by vast expanses of empty space, is meant to illustrate to the layperson that, contrary to what they previously believed, supposedly solid objects are mostly empty space. Eddington was far from the only one to engage in such talk. According to Reichenbach (1932, p. 22):

> the uniformly filled substance of the naïve picture exists no longer; instead, there is really nothing but tiny granules, which whirl past each other in violent motion. [...] Even the bridge pier of iron, which rises from the water and seems a symbol of repose and sustaining strength, reveals itself to the closer observer as a quivering structure, whose particles tremble in confusion, like the fine ramifications of a panicle of elderberries.

This is all well and good, but Eddington seems to be indulging in a lifting-the-veil conception of science when he invites us to reconsider ordinary beliefs; e.g. by claiming that a wooden plank upon which I step when entering a room is, in fact, not solid. This has been widely criticized – although of course to be criticized here is not the science *per se*, but the philosophical interpretation of it. Close to the mark was perhaps Stebbing (1937), who claimed that there was a danger in the intermingling between ordinary and scientific language; or, more specifically, in the use of ordinary expressions to convey scientific ideas.[2]

The problem with this kind of linguistic intermingling is that expressions in a natural language come with pre-packaged meanings which were settled way before theoretical investigation started. The conceptual core of solidity, for example, mostly had to do with impenetrability, or perhaps some kind of structural rigidity (e.g. something is said to be solid to the extent that it can resist penetrative force). More generally, solidity is a functional or dispositional characterization meant to

[2] Of course, to the extent in which Eddington was trying to divulge recent scientific results to a wider audience, which may or may not have possessed the conceptual tools to understand such results as they had been formulated, such an intermingling was somewhat unavoidable. But in such cases, one should preface such simplifications and analogies by disclaiming that they are loose talk, helpful tools for the uninitiated – but they should not be taken literally. During dissemination, we sometimes omit such disclaimers to preserve emotional impact (after all, one must be in a very impressionable state of mind to believe that stepping on a wooden plank is like stepping on a swarm of flies).

summarize important differences between solid objects on the one hand, and liquids and gases on the other. There can be, of course, a more scientific understanding of solidity which investigates the microstructural conditions underpinning an object's impenetrability and rigidity (in other words: an investigation on the so-called causal basis of solidity); and of course there could be a semantic dispute as to whether the meaning of "solidity" should be fixed by the ordinary or by the scientific understanding of the word, with corollary debates to accompany it (e.g.: can solidity be multiply realized?). But such disputes are no matter now; for, in these debates, the extension of the solidity-predicate is not at stake – it is, on the contrary, a given. But what we are discussing now is a far more radical proposal, one according to which the extension of the solidity-predicate is radically redrawn, perhaps by claiming, as Eddington may have implied in the passage, that nothing is solid. Something similar is going on in the case of the metaphysician who employs option (B); e.g., the extension of the predicate "is wrong" is usually taken to be non-empty; but if the moral anti-realist who supports option (B) is right, then it is empty.

So now the question is: can the extension of these predicates really be redrawn in the way suggested by the lifting-the-veil conception? There are some reasons to believe that doing so may be very problematic. This issue brushes against a monumentally complex problem about interpretational constraints: in its simplest version, the question is: is there "semantic glue to stick our words onto their referents" (Lewis, 1984, p. 221), and if there is, which is it? In our case, how do meanings (whatever they are) and therefore extensions, stick to predicates? How can we distinguish a good interpretation from a bad interpretation of "solidity", etc.. . .? Theories differ, but a baseline for interpretational constraint is what Lewis calls "Charity Constraint", according to which any candidate interpretation must fit with ordinary use of the term; this may not be a sufficient condition for selecting a correct interpretation, but it is a necessary one. A selected class of sentences of ordinary and current use, containing the expression, must come out as true for a candidate interpretation to be correct.

On a similar note, Eli Hirsch proposes an argument from interpretive charity in favour of the truth of ordinary sentences which may also be understood as providing metasemantic pressure in favour of ordinary interpretations: according to Hirsch, the correct use of the term is the ordinary one (Hirsch, 2013, p. 439). Korman (2020) offers a more detailed argument to this conclusion:

(K1) The most charitable interpretation of English is one on which "the wooden plank is solid" comes out as true
(K2) If so, "the wooden plank is solid" is true
(K3) If "the wooden plank is solid" is true, then the wooden plank is solid
(K4) So, the wooden plank is solid

The crucial points of discussions here are (K1) and the move from (K1) to (K2). As for the defence of (K1), a principle of interpretive charity must be deployed. Such a principle should maximize the speaker's rationality; when a fluent speaker of English utters a sentence like "the wooden plank is solid", nothing in their linguistic or extra-linguistic behaviour suggests that they are using the predicate "is solid" according to any other interpretation but the ordinary one, according to which the predicate serves to, e.g. separate impenetrable entities from liquids or gases. There is no presumption that any other interpretation is at play here. According to that interpretation, the sentence comes out as true.

But how to move from (K1) to (K2)? Here is where things become slightly more complicated. One may want to argue that a non-deductive argument is needed here. For example: there is no reason to think that rational and fluent speakers of English make that many systematic and inexplicable errors when employing the solidity-predicate; so we might inductively or abductively argue that that interpretation is the correct one. But we might deploy a stronger argument by adding an additional premise, and, more precisely, a premise concerning the kind of interpretive charity deployed for the job. Hirsch's principle of interpretive charity is meant to be *constitutive* of meaning, in the sense that sentences (or utterances perhaps) necessarily have the meaning assigned them by considerations of charity. According to Hirsch (2013, p. 438):

> [t]he principle of interpretive charity implies, roughly put, that it is metaphysically impossible for your assertions to be stupider, crazier, or less rational than your (actual and potential) linguistic behaviour dictates. You are in this respect innocent until proven guilty. This form of metaphysical grace is, according to the principle, constitutive of linguistic meaning.

Therefore, interpretative charity as semantically constitutive sanctions the shift from (K1) to (K2) and provides interpretive constraints based on charity in favour of ordinary interpretations.[3] Interpretive charity so understood is a widely shared principle, first and foremost by Davidson (1984); but even those who reject

[3] A crucial point about Hirsch's argument must be noted. The argument scheme of which (K1)–(K4) is an instance does not always produce true conclusions, because the premises are not always true. We circle back to what may be (or may have been) another ordinary belief, viz. that there is a Loch Ness monster. The following sentence is false:

(N1) The most charitable interpretation of English is one on which "there is a Loch Ness monster" comes out as true

That (K1) is true, while (N1) is false signals one important fact: considerations of charity must allow for the possibility of error. When someone, perhaps very gullible townsfolk utter "there is a Loch Ness monster", nothing in their linguistic or extra-linguistic behaviour indicates that they are using those expression but in the most ordinary sense; with "Loch Ness monster" they clearly mean a monstrous creature that lives in the lake, and with "there is" they clearly mean the same thing they mean when they claim "there is beer in the fridge". And yet, on that interpretation, the sentence comes out as false. According to Hirsch (2002), this has to do with the fact that charity maximizes *rationality* – not *truth*. The most charitable interpretation is one that makes the subject the most rational with respect to its linguistic and extra-linguistic behaviour; but that doesn't mean that their utterances are automatically true – for example, if some empirical data is still missing. Here is therefore a crucial difference between the job of the metaphysician and that of an explorer: the latter, but not the former, is involved in gathering new empirical data. This means that interpretive constraints based on ordinary use can shield ordinary beliefs from the metaphysician, but not the explorer.

interpretive charity as "constitutive" or "foundational" still take it to be a somewhat reliable principle; e.g. according to Pagin (2006): "it is a matter of nomic regularity that if an interpretation reliably lives up to the requirements of the principle of charity, by the interpreter's standards, then the interpretation is correct".

Roughly put, terms for which there is extensive and ordinary use in a pre-theoretical setting – viz., before the physicist or the metaphysicians come along– have their meanings settled through the linguistic interactions involved in such setting; which generates genuine pressure against any other interpretation of it. Perhaps a hidden assumption here is that meaning is constructed through the kind of successful communicative interaction that ultimately motivates the creation and adoption of language, rather than being the result of private conceptual activity. As noted in Dresner (2006, 2011) if the latter were the case, interpretive charity would be a less plausible principle.

Look at it this way: if one uses an expression with a deviant interpretation, either massive failures of communication ensue, or several "what I meant" disclaimers must be added to course-correct the discussion. E.g., suppose that I used "Joe Biden" to refer to Donald Trump; massive failures of communication would ensue; alternatively, I would have to add many disclaimers clarifying my utterances: e.g., "Joe Biden is not the president the United States, but this does not mean that the GOP has won the election: with 'Joe Biden' I simply mean Donald Trump". The supposedly shocking reveal ("Joe Biden is not the President of the United States") is uttered in the spirit of trickery, and, upon further investigation, does not in fact establish that Joe Biden is not the President of the United States. Either way, it would not be far-fetched to conclude that I am using "Joe Biden" incorrectly.

The same goes, I submit, for supporters of the lifting-the-veil conception, both in science and metaphysics; viz., Eddington and his (hypothetical) interpretation of the solidity-predicate, and the supporter of option (B) who takes the premises of Moorean Arguments to be false. For instance, Eddington's (hypothetical) reading of the solidity-predicate as having an empty extension would generate many failures of communications; after claiming that the wooden plank is not solid, any speaker would need to add several disclaimers whose entire purpose is to course-correct their previous utterance; e.g., they would have to add that no, this does not mean that the plank is not impenetrable, and no, this does not mean that the plank is a liquid, or a gas. The moral anti-realist sympathetic to the lifting-the-veil conception would need to similarly course-correct their utterances: e.g., "no, killing babies for fun is not wrong, but this does not mean that you are allowed to do it; by saying that 'is not wrong' what I meant was...". And the same goes for the mereological nihilist: "no, there is no tree in my backyard, but please bring a hammock anyway next Sunday; what I meant was...". These considerations above suggest that such terms are being used incorrectly by the supporter of the lifting-the-veil conception. As intuitively compelling as it may sound, to cash out what is merely the case as what is not the case (viz. as a kind of falsehood), as A_1–A_2 dictate, would lure the metaphysician of the anti-realist variety in bad linguistic practice, if not a veritable metasemantic error.

This line of reasoning may shine some light on the nature of the so-called "Moorean certainties" and their alleged unassailability. A Moorean certainty is not simply a sentence whose truth appears to be intuitive, or obvious. For metaphysicians are not bound to preserve what appears to be intuitive, or obvious (obvious to who, incidentally?). Metaphysicians, however, are speaking natural languages which have been devised and infused with meaning well before they came along. They are free to frame their discipline as an investigation of what is really the case, as opposed to what is merely the case (if they can figure out a way to make sense of the distinction – as we are trying to do); but what they cannot do is hijack and repurpose already existing expressions when telling us how things really are. Moorean certainties are, in a sense, the extra-metaphysical limits of metaphysics.

Relatedly, we are now able to recognize when a sentence is not a Moorean certainty. E.g. in the previous chapter, we have submitted that just as mereological nihilism can be threatened by a Moorean Argument starting from the premise "there's a tree in Lorenzo's backyard", mereological universalists can be threatened by a Moorean Argument starting from the premise "there are no cat-dogs"; but we have also seen that the two cases are not exactly speaking symmetrical. This asymmetry can now be articulated more explicitly. I have no reason to doubt that "there is a tree in Lorenzo's backyard" may be a Moorean certainty, so understood; but "there are no cat-dogs" is surely not one. The reason is straightforward: because "cat-dog" is a term introduced by the metaphysical discipline of mereology, there are no pre-theoretical and ordinary utterances of "there are not cat-dog"; in other words, there is no ordinary use of "cat-dog" that would sanction the truth of negative existentials about cat-dogs by considerations of interpretive charity. A little more grandiosely, the non-existence of cat-dogs is not part of "common-sense ontology" (whatever that is), for the simple reason that common sense does not deal in cat-dogs. In conclusion, "there are no cat-dogs" is merely a somewhat intuitive sentence (intuitive for some layperson, presumably), and the mereological universalist is more than free to dispute it.

3.2.2 Towards Compatibilism, and a Factive Conception of Appearances

The quoted passage from *Nature of the Physical World* invites the thought that Eddington supported a lifting-the-veil conception, according to which, although it would appear to be the case that a wooden plank is solid, this is not the case (as dictated by A_1–A_2). This stance has been shown to be problematic.

But perhaps we have been too hasty in our judgment; perhaps, on a more charitable reading of the above passage, Eddington could be interpreted as saying not that modern science reveals that objects like wooden planks are not solid, but that, while one would normally take an object's solidity to entail a *plenum*, or that the object is made up by dense or continuous "stuff", modern science demonstrates that this is not so. In short, what the physicist is inviting its audience to reconsider are not ordinary beliefs, but the entailments holding between such ordinary beliefs and other beliefs. This "entailment-reconsidering" is close in spirit to the strategy of compatibilism (C); as we have seen, e.g. the peculiarity of the mathematical nominalist who holds (C) is that they do not invite one to reconsider the ordinary belief that there is a prime number between 3 and 7; they are rather inviting to reconsider the entailment between that belief and other beliefs – e.g., in its most general formulation, the belief that mathematical nominalism is false. Strategy (A) and (B) agree that if there is a prime number between 3 and 7, then mathematical nominalism is false; option (C) disagrees.

There are plenty of ways to formulate "entailment-reconsidering" within a compatibilist perspective. E.g., the mereological nihilist may be intended, not as reconsidering the claim that there is a tree in Lorenzo's backyard, but as reconsidering the entailment from "there is a tree in Lorenzo's backyard" to "there are mereological fusions" (Contessa, 2014). Similarly, the moral anti-realist is not reconsidering the claim that killing babies for fun is wrong, but that killing babies for fun is wrong entails that, say, morality is real, or that "is wrong" is joint-carving, or some other formulation of moral realism.

This is where a factive conception of appearance, according to the A_3–A_4 reading of the distinction between reality and appearances, may be deployed to formulate this "entailment-reconsidering" in its greater degree of generality. According to this approach, matters of realism are matters of what is really and not just merely the case; therefore, when faced with a supposedly threatening ordinary belief or Moorean certainty "p", the right move for the compatibilist is not to invite their interlocutor to reconsider the belief that p, but to reconsider the entailment from "p" to "it is really the case that p" – which, given A_3–A_4, is poorly motivated.

This move, of course, raises some questions on its own; if we cannot use, as the supporter of option (B) does, a non-factive conception of appearances to entirely deflate the distinction between what is really the case and what is merely the case, then how can we treat it? This is what we now turn to.

3.3 Reality Is Not Layered

This section mostly comes in the form of a warning, formulated through a simple principle: thou shalt not reify appearances. Or, perhaps more precisely: thou shalt not *separately* reify appearances. This principle appears so obvious to me that I would not know how to motivate it, except through the broadest strokes. Eddington was surely right to claim that the world as it ordinarily appears to us, and as it is investigated by our theoretical endeavours are very different beasts; a wooden table as we encounter it in our everyday life is surely very different from a wooden table when studied in a microphysics lab – but it is not literally true that there are two tables, as Eddington may or may not have claimed. To reify appearance seems an instance of "metaphysical double vision", to borrow an expression from Mackie (1977, p. 366), in which we unduly crystallize different stages of theoretical investigation in *ersatz* realities of their own. The table is the same; but it is categorized in two different ways.

As suggested in the previous chapter, my main adversary in this section is a layered conception of reality as in Schaffer (2009), in which a domain of entities, be they facts or objects, is ordered by a relation of metaphysical grounding or relative fundamentality, from the less fundamental (more derivative) to the more fundamental (if not absolutely fundamental – if grounding is well-founded). Reality is to be found at the bottom, if there is one. This conception may *prima facie* sit well with our factive conception of appearances, in which Moorean certainties may correspond to facts located somewhere in the fundamentality hierarchy (e.g. for a moral anti-realist the fact that killing babies for fun is wrong will presumably be a very derivative fact).

On second thought, however, this conception of appearances, and their relation with reality, is not a good fit for us. Given the prominence of metaphysical grounding in recent literature, doubt and skepticism takes many forms, from more localized worries to widespread doubts concerning the very intelligibility of grounding notions (Wilson, 2014; Maurin, 2019). But before I present my worries, I must be very clear about what it is that I am arguing against: that I find the fundamentality-based layering of reality unacceptable does not entail that I find all manners of layering unacceptable. There are perfectly acceptable dimensional/mereological layerings; similarly, there are scientific layerings of the world that are only indirectly correlated with what is "metaphysically fundamental", whatever that means (Bhogal & Perry, 2017; Hicks & Schaffer, 2017). Not only that: even within metaphysics, it is not all conceptions of ungroundedness and fundamentality that trouble me, and that thus will need to be decoupled from our discussion. In what follows, with "layers" and "layering" I will mean fundamentality-based layerings, used to distinguish what is "fundamentally the case" to cash out the distinction between reality and appearance; in other words, I have trouble with the idea that we should look at the bottom of a grounding-generated hierarchy of relative fundamentality to decide what is real, and what is mere appearances. This is a non-vacuous target, albeit not an all-encompassing one either. It is non-vacuous because grounding and fundamentality have patently been linked to matters of realism, in a thread going from Fine (2001) to Schaffer (2009), and with undertones bleeding all over the debate (the ever-present metaphor of what is fundamental as "what God needed to create when making the world" at least suggests that what is fundamental, unlike the derivative, is what is *real* – viz. is what our metaphysical and ontological theories are committed to as non-redundant characterizations of a mind-independent reality); but it is not all-encompassing either, as there are conceptions of fundamentality that do not overlap issues of realism and anti-realism. These also will need to be decoupled from our discussion. As a matter of fact, as Fine (2001, p. 22–24, 27–28) points out, matters of fundamentality/ungroundedness may sometimes not perfectly align with matters of reality, and someone might indeed follow in his lead in their understanding of a grounding-based layering as entirely within "equally real portion of reality" (perhaps the entire point of this section can be summarized as arguing that such an expression is meaningless).[4] Relatedly, as we will see at length in the course of this

[4]Amongst others, one might doubt that Schaffer's (2010) priority monism is in the business of claiming that only the whole – viz. the entire cosmos – is real.

book, there are plenty of non-causal dependence and explanatory relations that I deem to be perfectly acceptable (although some of them I take with an irrealist spirit), so even thought I reject the idea of a world layered from the more real to the less real, there are plenty of ways for things to non-causally depend or be explained by other things.

With this clarification in mind, my worries are threefold. Firstly, the picture I sketched below requires both reality and appearance to be parsed into discrete units such as facts or objects, which, as we have seen in the last chapter, is not something which sits well with our set-up: for it is not required to think of perspicuous nor merely true sentences to "carve up" reality in this fashion. This first worry is easily survivable by switching from a predicative to an operationalist understanding of grounding, so as to keep neutral on such matters.[5] On the predicative conception of grounding, grounding statements will take the form of, say: [p] is grounded by [q], where [. . .] is a device that turns sentences into names for facts, and "is grounded by" is a binary predicate; but on the operationalist conception, a grounding statement may take the form of, say: p because q; in this case, sentences are used directly, flanked by a sentential connective "because"; on this second conception, facts are neither referenced nor quantified upon in grounding statements.[6]

My second worry is not as treatable: this conception violates my guiding principle for the section, given that they (separately) reify appearances; furthermore, they reify appearances in a way which is potentially mysterious, and not very appealing. Metaphysical grounding, so understood, is a way to take very seriously Fine's (2001) distinction between reality and appearance, and crystallize into a distinction between "real" entities and other entities; these may be fundamental objects or objects, or, on an operationalist perspective, fundamental matter-of-facts.[7] The corollary always stands that, on this account, only some things (or matters-of-fact) are real; this straightforwardly entails that, while some entities enjoy the kind of privilege that allows them in the very exclusive club of "real" entities, others are but "shadowy [. . .] semi-existents" (Cameron, 2008, p. 7). What the privilege of real entities ultimately consists in, remains to be seen. Consider this question: is this relation of metaphysical grounding or relative fundamentality supposed to be internal or external? Neither option is ideal. If it is external, then the only difference between the more fundamental and the less fundamental – and thus the entire distinction between reality and appearances – would amount to the imposition of an external relation on what would otherwise be a class of indistinguishable entities; e.g., in the case of mereological nihilism, a mereological atom and the tree in their

[5]The same move is possible, as hinted in the previous chapter, for truthmaking as well – thus this discussion also applies, *ceteris paribus*, to deflated truthmaking.

[6]For an introduction on the logical form of grounding, see Trogdon (2013) and Trogdon and Bliss (2021).

[7]That such matters-of-facts are not, on the operationalist perspective, reified into heavyweight facts, does not mean that we cannot express the separation between more fundamental and less fundamental matters-of-facts; for this purpose we can use, as I will do at length in the course of this book, a notion of generalized identity. I leave this exercise for another time.

backyard would be metaphysically on a par when considered by themselves; the distinction between them only arises by taking their relations into account. If, on the other hand, this relation is internal, that would entail that there is something about the relata which makes them suitable to be so related; in other words, there is something about certain entities which makes them suitable to be more or less fundamental than others. But this only raises the question: what kind of intrinsic metaphysical privilege do "real" entities enjoy which is denied to the others? This is where we encounter cryptic remarks about fundamental entities "transmitting their being" to the derivative ones, or, conversely, about derivative entities "borrowing reality" from the fundamental ones – which of course raise more questions than they answer.[8] More generally, to reify the distinction between appearance and reality by claiming that some entities have a mysteriously decreased metaphysical status seems more mysterious than clarifying; and the suspicion might creep in that to layer reality in this fashion is to impose upon the world what would otherwise merely be the direction of our inquiry.

Supporters of these conceptions occasionally claim that derivative entities or matter-of-facts do not constitute a genuine ontological inflation, and they are not "an addition to being" (Schaffer, 2009, p. 353). But this is not an answer, just another way to raise the question. The point remains that, on these accounts, not everything is fundamental, or real: so the question also remains as to why the derivative is "no addition to being" and thus does not deserve the metaphysician's respect.[9]

[8] There seems to be some agreement that metaphysical grounding is internal (for a critical take on the subject, see Litland, 2015); in fact, according to Bennett (2011, p. 32) grounding is super-internal, in the sense that the intrinsic nature of the grounding entity is sufficient to establish the existence of the grounded and the holding of the relation between them. In that case, a whole lot more needs to be said about the nature of the supposedly "real" entities which is ultimately responsible for the whole real/apparent infrastructure. Incidentally, this topic is related to a very important discussion concerning meta-grounding or meta-explanatory questions, which I will discuss in Chap. 8.

[9] My worry loosely resembles Oliver's (1996, p. 31, fn. 30) doubts with Armstrong's (1996, p. 12) famous claim that "[w]hat supervenes is no addition to being": "[s]ince supervenient entities exist and are not identical to the entities upon which they supervene, they must be an ontological addition." The crucial problem there was that the supervenient and its base are numerically distinct, thus making Armstrong's claim poorly motivated. A similar dilemma has in fact been proposed for grounding as well, which is occasionally taken to entail reduction-as-identity (Rosen, 2010). E.g. Audi (2012, p. 110) claimed that, to the extent in which reduction, or nothing-over-and-aboveness (a close cousin of Armstrong's no-addition-to-beingness) has to be cashed out in terms of identity, and to the extent that grounding is asymmetric, the grounded is in fact something over and above what is doing the grounding. So one might ask: why does the grounded not deserve the metaphysician's respect? And, similarly any declaration that a grounded entity which is numerically different from its grounding base is still nothing over and above that base incurs in the same problem; see Trogdon's (2013, p. 113–114) discussion on Schaffer's proposal. Some options have it better in this regard; e.g. Azzouni's (2012) take on grounding according to which derivative entities do not effectively exist. Solutions which radically revise the logical form and/or formal properties of grounding (Jenkins, 2011) come closer to positions I will espouse in this book. More below.

Finally, a third worry. On this conception, it is not just appearances which are reified: the link between reality and appearance also is. A layered conception of reality is not simply problematic because it consists in layers, but because it also consists *in a layering*. But why is it problematic to enrich reality with a relation of metaphysical grounding or relative fundamentality? Firstly, it seems to be again a result of metaphysical double vision, the reification of a mere direction of enquiry; to use a trite metaphor for expressing matters of realism, it doesn't seem to me that when God was creating the world, They needed to create grounding as well. But then again, I do not presume to know what God needed or needn't do; secondly, and more to the point, the reality of grounding would have pernicious consequences on the methodology of metaphysics; as humorously pointed out by Dasgupta (2017, p. 74), "[t]he job of a metaphysician, on this new conception, [is] to peer into reality and discern where these 'groundons' were flowing (of course, to see these groundons one needed goggles provided by specialist departments)."

Incidentally, this is not solved by a switch to operationalism: one of the lessons learned in the previous chapter is we can meaningfully ask about the reality of something irrespectively of the syntactic category used to express it; so, the question as to whether grounding is a genuine aspect of reality can be asked irrespectively as to whether grounding is expressed by a predicate or an operator. There is no refuge to be found in syntax: for both expressions, we can ask whether they are joint-carving: viz., we can equally ask whether $\mathbb{S}$(is grounded by) or whether $\mathbb{S}$(because). Therefore, although Dasgupta's worries about the reality of grounding may very well legitimate (in fact, they are my worries a well), such worries cannot be side-stepped simply by tinkering with the logical form. Pretending to resolve the question of the reality of grounding by appealing to a deflated notion of grounding is to indulge to the Ontological Temptation: viz. to think that discussing the reality of grounding amounts to discussing the metaphysical status of an entity (in this case, the existence of a relation expressed by a binary predicate). More to the point, it involves the Ontological Fallacy of thinking that because the right ontology is not present (allegedly, operators don't correspond to entities as predicates do), then the question of realism of grounding is settled in the negative. This approach is compounded by the further mistake that it is not exactly clear whether operators do not, in fact, correspond to entities as predicates do: as we have seen in the previous chapter, if the grip of the Ontological Temptation is strong enough, someone might very well claim that they do. This is not to say that I believe grounding to be real, viz. that this layering of reality is real. I very much doubt it is. But this cannot be established by way of trickery concerning predicates or operators.

By this point, the question of the reality of grounding is very much open; I would make the further claim that if grounding is to be used to distinguish reality from appearance, then this question has to be answered in the affirmative. Here is where I risk saying something controversial again: if the grounding predicate-slash-operator is used to draw a distinction between reality and appearance, then it will necessarily be joint-carving: for how could grounding successfully interface reality and appearance, if it itself was not real? E.g., if this deflated conception of grounding was allowed to be "relative to certain interests that might vary from culture to culture or time to time" (Dasgupta, 2017, p. 78), then the distinction between reality and appearance could be just as volatile, which is patently absurd.

Let me make something clear: I have claimed that Moorean certainties are sentences for which there is metasemantic pressure, ultimately based on ordinary use, to take them as true. But of course the ordinary use of an expression, within languages or across languages, wildly varies between linguistic communities; thus it may very well be that Moorean certainties vary across such linguistic communities. More generally: there is no unique realm of appearances, but many (so to speak). This is more of a feature than a bug of the account, for the world as it appears to us is most certainly not the same as it appears to far away linguistic communities. However, the same does not hold for reality: metaphysical realism, in its broadest interpretation according to which there is a reality independent from our linguistic or conceptual choices, forces our hand by dictating that while realms of appearances may be many, reality is only one. And so, if reality is to be found at the bottom of the grounding hierarchy, viz. what is absolutely fundamental, then this bottom had better be the same for everyone. If, on the contrary, grounding is to be freed from the shackles of realism and objectivity in the way Dasgupta suggests, then we cannot use it to frame the distinction between reality and appearance (and, to be clear, this is not the use Dasgupta himself has for it). Reality cannot be the spot where people stop asking grounding questions, simply because different people stop asking questions in different spots.

In conclusion, if grounding is to be used to frame the distinction between reality and appearance in a factive conception of appearances, it will need to be to some extent real itself – irrespectively as to whether it is expressed by a predicate or a connective. However, just as I think that reality does not encompass a mysteriously diminished realm of appearances, I do not think that reality encompasses such as thing as a reality-layering relation of grounding, viz. patterns of an out-there relation linking what is really the case with what is merely the case. Grounding, of any kind, is not the right tool for the task at hand.

3.4 Appearances, Factuality, and Truth

So, I will not reify appearances, and, in fact, my approach to this potentially mysterious realm of appearances is entirely deflationary – and entirely based on the resources provided by my account so far.

3.4.1 A Deflationary Account of Appearances

I have advocated for a structure-based compatibilist solution to Moorean Arguments; such a strategy cannot simply rely on the claim that reality exhibits structure: one also needs the assumptions that sentences may be true without revealing such structure. Namely, there can be mere truths, which are not metaphysically perspicuous (this amounts to the rejection of Rayo's Metaphysicalism^S). This approach suggested a natural way to understand Fine's distinction between reality and appearance, which is now understood as the distinction between perspicuous sentences and mere truths. We begin with the claim:

A_5) it appears to be the case that ϕ iff "ϕ" is true and if it is a Moorean certainty that ϕ

(A_5), as an elaboration on a factive notion of appearances, entirely exhaust the nature of appearances – it's in accordance with (A_3). As we have seen: whether something counts as a Moorean certainty depends on metasemantic considerations; viz. that there is pressure, based on ordinary use of the relevant expressions, to take the sentence as true. Thus if a sentence s_1 is a true Moorean certainty, there is no special metaphysical status assigned to the fact that s_1, nor any special semantic status assigned to "s_1". Appearances are simply truths, formulated in a language which is devised and infused with meaning for what may very well be practical purposes (data sharing, action coordination, etc.. . .), which are not necessarily that of conveying any information about the structure of reality[10]; any truth, mere or otherwise, expresses what is the case, in the sense that it conveys factuality (by saying that truths convey factuality, I patently do not mean that truths correspond, or are made true, by heavy-weight facts: on my account, there are no heavy-weight facts. I will come back later to the notion of factuality).

To the distinction between *truths* and *mere truths* correspond the distinction between *appearances* and *mere appearances*. From the starting point that certain things are the case (e.g., there is a prime number between 3 and 7), metaphysicians can then disagree as to whether that is really or merely the case. As we know by now, this is the difference between perspicuous truths and mere truths. The following (A_6) exhausts what we may call *mere appearances*:

(A_6) it merely appears to be the case that ϕ iff it appears to be the case that ϕ and "ϕ" is merely true.

(A_6) is just as deflationary as (A_5): mere truths do not carve out any ancillary "realm of mere appearance" in any reified form, such as the one we found in the layered conceptions of reality which deployed metaphysical grounding. And therefore, there is a key difference between my account and Lipman's (2018) framing of factive conception of appearances: I do not claim that certain matters of fact enjoy a privileged as opposed to a non-privileged status (e.g., different kinds of existence, fundamentality, etc.. . .). My way of framing this distinction is through language; certain pieces of languages are perspicuous as opposed to non-perspicuous – but there is one reality, of one kind, which can be described either perspicuously or non-perspicuously.

[10] Again, see Rayo (2013, p. 10–11).

To summarize my deflationary account to appearances, the realm of appearance is the realm of factuality. Simply put, it's what is the case. It is the case that Paris is the capital of France, and that there is a tree in Lorenzo's backyard, but it is not the case that there is monster hiding in the depth of Loch Ness. It is also the case that killing babies for fun is wrong. However, for a moral anti-realist "killing babies for fun is wrong" is a mere truth, and thus a mere appearance. This sentence does not correspond to a less fundamental fact; the anti-realist position has to be understood negatively, simply as claiming that "killing babies for fun is wrong" and other moral judgments do not reveal any degree of structure in objective reality, presumably because ~𝕊(is wrong) and ~𝕊(is right). Thus, the distinction between reality and mere appearances is deflated with the help of the distinction between perspicuous truths and mere truths.

3.4.2 About Truth

At this point, I should say something on the nature of truth and its value for metaphysics. The idea behind the factive conception of appearances is that the metaphysician's struggle to distinguish reality from appearances ought not to be conducted at the level of truth vs. falsity, and that the increased metaphysical value of a subset of truths is to be cashed out by a properly metaphysical notion (in my case, metaphysical perspicuity). This is not a trailblazing stance, of course: it is but the latest version of Fine's (2001) seminal reconceptualization of anti-realism in metaphysics: to borrow a slogan from Finocchiaro (2021), "truth is not enough". Truth is not completely irrelevant for the purpose of metaphysics (after all, I endorse the Factivity of Perspicuity: perspicuity is a subset of truth), but most of the times it is not sufficient for the metaphysician's purposes.

What is truth, then? This background already imposes some constraints on what we take truth to be: many versions of the correspondence theory of truth are excluded by the account (at least the more robust ones, such as the notion of correspondence-as-isomorphism), and truthmaking theory can only be accepted in its most deflated version, without reified truthmakers.

Truth goes hand in hand with factuality, the two being inextricably linked by Tarski's disquotation schema. To claim, as above, that the realm of appearance is the realm of truth, is just another way of saying that appearances are simply what is the case (and, of course, we don't need metaphysicians to know what is the case). But this association can also be at the root of some difficulty concerning the understanding of truth in this framework, in which truth and thus factual non-perspicuous sentences exist. This may strike the reader as rather mysterious: if there is indeed structure, how can a sentence convey factuality without at the same time conveying structure? This is a broader discussion which can only be partially undertaken here; yet something needs to be said about the nature of truth and factuality, its relation with perspicuity, and its role in metaphysical methodology at large.

I will begin by addressing what I believe to be the roots of the worry concerning the impossibility of non-perspicuous true sentences. This worry appears to be based on two assumptions. The first assumption is, again, that reality has a structure in the ontological sense that it is parsed into discrete units, such as objects, properties, facts, or what have you. The second assumption is that a true sentence has truth-conditions, and that such truth-conditions are understood in an inherently metaphysical way. What do I mean by this? The idea is that truth-conditions provide the conditions for the truth of the sentence, in the sense that they articulate what is required from reality for the sentence to be true; in other words, the semantic meta-language is providing information about the world's structure. In this sense, if ordinary sentence such as "the ball is red" or "there is a tree in Lorenzo's backyard" are true, then the semantic meta-language references and/or quantifies over some of these discrete units in which reality is parsed into. And so, there is really no way for "the ball is red" to be true, without the supposedly red ball being one of the denizens of reality, and the same for "there is a tree in Lorenzo's backyard" and the tree in Lorenzo's backyard. The result is a very naïve picture of the interface between language and reality which does not really allow for the separation between truth and perspicuity.

This picture is naïve simply because its assumptions are naïve. As for the first one, we have seen how this "ontological" picture of reality has been superseded by this account by now; but let us assume it for the sake of the argument. The real culprit here is, I take it, the excessively metaphysical reading of truth-conditions. After all, when it comes to Tarskian truth-conditions, they were first proposed and adopted not because they somehow allowed any special insight on the nature of reality (whatever that means), but because they were compositional. Their compositionality, and resulting material adequacy, is a very useful tool when understanding the notion of truth, but any further metaphysical reading is threading on dangerous grounds; after all, it is the job of truth-conditions to offer virtuous biconditionals with "... is true" on the one side, and the conditions itself on the other. They function as useful bridge principles between two classes of sentences. But it is a further metaphysical claim, and one not very well motivated, that one of the two sets of sentences gets a grip on reality better than the other. In other words, it is far from clear that the metalanguage of Tarskian truth-condition is metaphysically perspicuous. Consider this: what is required from reality for the truth of "it's storming" is presumably the same than what is required for the truth of "there is a storm". But if we do not have the further reassurance that Tarskian truth-conditions for "there is a storm" get a better grip on reality that those for "it's storming", there is no certainty that Tarskian truth-conditions of "there is a storm", with the storm-entity they involve, will gets us closer to reality. This reassurance ultimately amounts to a claim about which sentences are metaphysically perspicuous: in short, it is a substantively metaphysical claim.[11] Devitt (1996, p. 163) may be sharing similar worries when discussing the topic of ontological commitment:

[11] Anti-realists about structure will perhaps claim that, metaphysically speaking, there is no difference between "it's storming" and "there is a storm", and that object-talk is only employed for its semantic virtues; see Rayo (2013, p. 13).

> I don't claim that there is no problem in establishing a commitment in the object-language, but simply that any problem that there is arises as much in the metalanguage. [. . .] Language does not suddenly become kosher when you start doing semantics. The idea that talk about the world is unclear and in special need of interpretation, yet talk about language and its relation to the world is straightforward on the face of it, reflects the damage of years of living under the linguistic turn.

This is not say, mind you, that semantics can never be metaphysically illuminating. Indeed, that would be the job of what Sider has called "metaphysical semantics". The job of a metaphysical semantics is exactly that of explaining what a true sentence, of any kind, tells us about the world. After all, perhaps objective reality does not comprise a special storm-entity, but if it is true, "there is a storm" must be telling us something about reality – even though it is not *prima facie* clear what it is. In Sider's (2011, p. 133) terms, metaphysical semantics, by "giving meanings in fundamental terms, it seeks to achieve something not sought by linguistic semantics: to show how what we say fits into fundamental reality." As one can imagine, this is an overly ambitious project, which has not yet left the programmatic stage.

In a way, this naïve picture of the interface between language and reality, and the ensuing collapse between truth and perspicuity – viz. a wholesale adoption of Metaphysicalism^{S}–, is at the root of the layered conception of reality in the first place. For it is only on this very naïve picture that the apparent truths of certain sentences (from ordinary macroscopic discourse, special sciences, etc.. . .) reveal the presence of supposedly distinct levels of reality, furnished with their own objects, properties, relations, and facts (Heil, 2003, p. 58). Therefore, a simple yet effective way to reject such layering consists in rejecting the two naïve assumptions that underlie it: first, by reiterating that there can be truth without perspicuity; secondly, by pointing out that a sentence's perspicuity is not necessarily a matter of it parsing reality into discrete units such as facts, objects, or what have you: to the extent in which matters of perspicuity are understood as matters of realism, this amounts to an overindulgence in the Ontological Temptation.

Finally, one may wonder what kind of positive theory of truth can be offered to complement my account of truth and perspicuity. This mostly falls beyond the scope of this book, but there are two interest points, and potential directions for future research, which can be made in this regard. Firstly, based on some of the considerations previously put forward in this chapter, someone may find a deflationary theory of truth appealing. The idea behind a deflationist theory of truth is, to put it bluntly, that there is no deep nature about truth. All that we are saying when claiming that, say, "killing babies for fun is wrong" is true, is just that killing babies for fun is wrong: we are simply injecting the content of the sentence in object-language discourse, and we are not necessarily engaging in any conversation about the structure of reality – viz., whether there is a moral aspect to reality, or whether $\mathbb{S}$(is wrong). Some of the considerations put forward in this chapter suggest this interpretation: natural languages developed independently from our scientific or metaphysical investigations, and so their function may not necessarily map anything of value for them. Their value lies elsewhere, mostly its utility for data sharing, action coordination and so forth. And so, if the value of language lies in its communicative purpose, the value of truth may be just that of disquotation – viz. to reaffirm that a sentence may indeed be *used* to ensure successful communication and coordination.

The simplest interpretation of this point, viz. that a true sentence is simply a useful one, is of course untenable. For intuitively a true sentence, at least a declarative true sentence, is also *factual* in the sense that it expresses some matter-of-fact about a specific topic; in other words, a declarative true sentence expresses what is the case. I wouldn't know how else to express or clarify this notion without compromising myself one way or the other: again, we do not want to say that declarative true sentences correspond to, or are made true by, facts or facts stand-ins. More generally, we should not understand factuality of a sentence as its inherent aptness to recapture whatever degree of worldly complexity. After all, that is what perspicuity is for.

I am not sure about how to articulate a metaphysical backdrop for factuality without indulging in some ontology, and in fact I am not even sure that it is possible in the first place.[12] Yet I remain convinced that there is a difference between a certain domain of discourse, understood as a collection of sentences, being *factual*, and it being *perspicuous*. For it is one thing to claim that there are objective matter-of-facts concerning such discourse, and another thing entirely to claim that it is real, or part of the structure of reality (e.g., it is a matter of fact that Milan is in Italy; but many metaphysicians would understandably balk at the notion of $\mathbb{S}$(Milan), or $\mathbb{S}$(Italy)). More to the point, two true sentences, one perspicuous and the other not, may both be expressing the same matter-of-facts, yet one of them may be expressing it better than the other. To use Fine's terminology, only one of them is properly expressing what is really the case, but not the other; but because we are refusing to indulge the Ontological Temptation, we must reframe ourselves from taking this sameness of matters-of-fact literally as implying fact-identity. We have understood in the previous chapter what it takes for a true sentence to be *perspicuous*; yet, we have not said much about what makes a merely true sentence *factual*. What is factuality, this "accord with reality" that some domain of discourse objective – assuming that it is not fact-correspondence, component-matching, or perspicuity of any kind?

As in the case of perspicuity, at some point we must stop in our enquiry. I suspect that, for our current purposes, primitive factuality is a good place to stop. Namely, we cannot expect to clarify in simplest terms what it is for a true sentence to express what is the case; whereas we can say something more about a certain sentence being perspicuous.

[12] For a laudable effort, see Devitt (1996).

3.5 The Interface Between Reality and (Mere) Appearances

Now that we have a framework for both reality and appearances, it is time to say something about the interface between them. This will finally be an opportunity to say something about the methodology of metaphysics.

At it is often the case in philosophy, this interface between reality and mere appearance involves a way to bridge two distinct classes of sentences – the former allegedly more problematic, and the latter allegedly more secure (from the point of view of the metaphysician, at least). In the context of my approach, the kind of threat posed by these allegedly problematic sentences does not involve the ontological commitments of sentences, nor the naturalness or fundamentality of certain entities: rather, the metaphysical "danger" posed by a discourse is its non-perspicuity; so the idea is that we might want to interface reality and appearances by providing some kind of bridge between perspicuous and non-perspicuous sentences.

Before we begin, a small complication. Keep in mind that perspicuity is cashed out as the joint-carvingness of its subsentential components; therefore, because $\mathbb{S}$ can be applied to expressions of any syntactic category in any sentence, of arbitrary logical complexity, we may conceive the distinction between perspicuous and non-perspicuous as the limiting case of a graded distinction between the *more* perspicuous and the *less* perspicuous. E.g., if $\mathbb{S}(\vee)$, $\mathbb{S}(p_1)$, $\mathbb{S}(p_2)$, $\sim\mathbb{S}(p_3)$, and $\sim\mathbb{S}(p_4)$, then "$p_1 \vee p_3$" will be less perspicuous than "$p_1 \vee p_2$" but more perspicuous than "$p_3 \vee p_4$". More perspicuous sentences, as one would image, reveal more structure than less perspicuous sentences, in the simple sense that it reveals more patterns of objective similarities holding between certain "facts" and others. If so, we might understand what is really the case as what is expressed by absolutely perspicuous sentences. Relatedly, I will occasionally talk about "the perspicuity of ϕ-talk", where ϕ stands for a subsentential expression: with this I mean that of two true sentences s_1 and s_2, built with the subsentential expressions, with the exception of ϕ, such that ϕ appears in s_1 but not in s_2, it is the case that s_1 is *just as or more perspicuous* than s_2, given that $\mathbb{S}(\phi)$.

3.5.1 About Explanation

Philosophers have used a variety of notions to offer bridge-principles such as the ones we need today: identity, reduction, supervenience, grounding, etc.. . .; I will not discuss the virtues or shortcomings of such notions, most of which are metaphysical artifices. Rather, I would like to start with a notion on which we have some intuitive grasp: explanation. *Prima facie*, it is simple enough to understand how what is perspicuous might explain what is non-perspicuous. Consider Sider's well-known example of the red-blue surface:

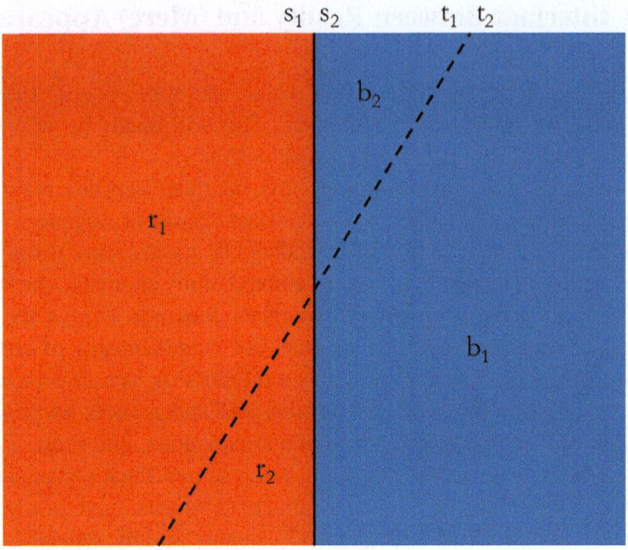

To clarify: s_1 is the surface at the left of the continuous line, composed of r_1 and r_2; s_2 is the surface at the right of the continuous line, composed of b_1 and b_2; t_1 is the surface at the left of the jagged line, composed of r_1 and b_2: finally, t_2 is the surface at the right of the jagged line, composed of b_1 and r_2. While we normally conceptualize the overall surface as being divided in a red half (s_1) and blue half (s_2), we might encounter an alien tribe which conceptualizes it as being divided in a "bled" half (t_1), and a "rue" half (t_2). After some initial confusion, we might be offered an explanation of the initially mysterious utterances of their speakers; e.g.:

e1) t_1 is bled because r_1 is red and b_2 is blue, or
e2) t_2 is rue because b_1 is blue and r_2 is red

These two utterances may provide an explanation in the simple sense that they make us understand some *prima facie* obscure claims, such as "t_1 is bled" and "t_2 is rue", in this case by bridging applicability conditions of familiar and unfamiliar colour-predicates. (e1) and (e2), are in this sense, illuminating.[13] What I am deploying is an ordinary non-causal sense of explanation. As such, explanation has unclear boundaries: there admittedly are many kinds of non-causal explanations,[14] but it is unclear whether by simply using the locution "because" under such an ordinary understanding, we are really picking up anything; this is, in short, an issue of intelligibility. In this regard, I prefer a tentatively optimistic and instrumentalist approach towards explanation, perhaps not dissimilar from Rosen's (2010)

[13]Wouldn't the members of the alien tribe find other supposed "explanations" illuminating? Absolutely. I will come back to this point shortly.

[14]For a recent investigation, see Lange (2016).

approach to metaphysical grounding: the notion can be useful, and as we will see, we can put some meat on these bones, even from such a sparse basis. We can always retreat, of course: whenever we feel the connection with our ordinary understanding of explanation is slipping, we can still take a step back. If not, let us see where it can take us.

I will employ an operationalist notion of explanation, so as to keep neutral on the metaphysical status of *explanans* and *explanandum*. I will characterize a family of non-causal explanations with a "...because..." sentential connective; more precisely, a binary operator with a plural operand with the following (very simple) logical form:

$$\text{p because } \Gamma,$$

where p is a sentence, Γ a plurality of sentences, and the "because" operator expresses a non-causal form of explanation. Some of the usual formal properties of this notion may be expressed with both singular and plural quantification over sentences, and by using "$\prec$" to express a one-many "...is one of the..." predicate of plural logic. If p because Γ, we might therefore say that p is jointly explained by all the facts expressed by the plurality Γ of sentences (again, let us not harp on the word "fact" too much – given that we are employing an operationalist conception, this is merely a figure of speech); then, the fact that p is partially explained by the fact that q, for every pΓ such that p because Γ. Thus, we might introduce a one-to-one notion of partial explanation as follows: (schematically, with the help of plural-quantification over sentences).

$$\text{p because}_{\text{p}}\ \text{q} =_{\text{df}} \exists\Gamma(\text{q}\prec\Gamma\ \&\ \text{p because } \Gamma)$$

Allowing pluralities of one, because$_{\text{p}}$ is a limiting case of because. For simplicity, we might express some of the formal properties of explanation focusing on the one-to-one explanation of because$_{\text{p}}$ instead of the one-to-many notion of because.

As it turns out, several of the usual formal features of explanation can be motivated by an ordinary understanding of the notion (Maurin, 2019). It is because there is something deeply unsettling about circular explanations that we might take because$_{\text{p}}$ to be asymmetric, and since all explanations must be to some extent informative, because$_{\text{p}}$ can also be taken to be irreflexive (formally, asymmetry entails irreflexivity, self-explanations being a limiting case of circular explanations – so perhaps the line of reasoning is that all circular explanations are problematic to the extent that they are ultimately uninformative); furthermore, because explanations do not survive adding irrelevant elements in the *explanans*, it is also commonly assumed that because$_{\text{p}}$ is non-monotonic. Transitivity is a bit more controversial: to the extent in which an explanation should increase understanding, chaining explanations through transitivity will at some point could very likely produce an unintelligible explanation. All these properties of explanation have been questioned sooner or later in the literature; yet in what follows, I will tentatively assume that

$because_p$ is asymmetric (thus irreflexive) and non-monotonic; that will provide a minimal formal grasp on the notion, as tentative as it might be.[15]

The next obvious question is: $\mathbb{S}$(because)? $\mathbb{S}(because_p)$? As I take "because" to stand for an entire family of explanations, the right answer is: it depends. In our study of Dispositional Reality, we will present and investigate various kinds of non-causal explanations; some of which I will take to be real, and others which I do not. But there's something we can say right now: the kind of explanation employed in (e1)-(e2) is what we may call *constitutive explanation* (Dasgupta, 2017), which I take to be non-real for the same reason I take metaphysical grounding to be non-real: because I do not wish to have a layered conception of reality, with red things in one layer, and rue things in another. They are the *same* things, just conceptualized differently.[16] This kind of explanation, in the broad and ordinary understanding that we tentatively used so far, is primarily an epistemic notion: an explanation supposedly provides a "a-ha!" sense of understanding in the enquiring agent (Woodward, 2018, p. 118–119); as such, I am entirely on board with Dasgupta's suggestion that explanation displays importantly social-relative, or even person-relative elements; whether something counts as an explanation may very well depend on interests and cultural sensitivities, just as (e1) and (e2) being illuminating (for us) on the ground that we are unaccustomed to the use of colour-predicates such as "rue" or "bled".

To clarify this point, here is a useful taxonomy regarding explanation: on the one hand, one might endorse what we may call "explanatory realism", described in Kim (2010, p. 149) as the position according to which

> C is an *explanans* for E in virtue of the fact that c bears to e some determinate objective relation R. Let us call R, whatever it is, an "explanatory relation." (The *explanans* relation relates propositions or statements; the explanatory relation relates events or facts in the world.) The explanatory relation is an objective relation among events that, as we might say, "ground" the *explanans* relation, and constitutes its "objective correlate." On the realist view, our explanations are "correct" or "true" if they depict these relations correctly, just as our propositions or beliefs are true if they correctly depict objective facts.[17]

[15] Although it is not always easy to separate discussions of (certain) non-causal explanations from discussions of metaphysical grounding, in favour of the asymmetry of explanation see Raven (2013, p. 193–194); against it, Bliss (2014) and Thompson (2016); for irreflexivity, see Trogdon (2013, p. 106), and against it Jenkins (2011); for transitivity, Raven (2013, p. 193–194), and against it Schaffer (2012). Rodriguez-Pereyra (2015) argues against all three. In favour of non-monotonicity, see Rosen (2010, p. 116), Trogdon (2013, p. 109), and Dasgupta (2014a, p. 4). For an up-to-date overview of this topic, Maurin (2019).

[16] Again, to anticipate things from later chapters, I take the only out-there component of constitutive explanation is generalized identity.

[17] The reader may notice that I have presented two criteria for realism about explanation: the first one, based on joint-carvingness and perspicuity; and the second one, based on Kim's passage. I don't think that they are in competition; they are rather complementary. Roughly put, the existence of an "objective correlate" to explanation, in Kim's words, may very well amount to the presence of a pattern of objective similarities signalled by the use of a "because" expression, if we assume that $\mathbb{S}$(because).

On the other hand, we might endorse "explanatory irrealism", or "explanatory internalism", according to which:

> the relation of being an *explanans* for, as it relates C and E within our epistemic corpus, is not, and need not be, "grounded" in any objective relation between events c and e. It is solely a matter of some "internal" relationship between items of knowledge.

If we choose to be explanatory realists, we still must clarify which relation exists between explanation and grounding: either tracking ("separatism") or identity ("unionism") – see Thompson (2016) and Maurin (2019) for the latter distinction. That said, when it comes to constitutive explanation, I squarely belong in the irrealist or internalist camp, as I do not accept the existence of any layering in reality for an explanatory relation to track, or to be identical to: if instances of because and $because_p$ hold in virtue of anything, they hold in virtue of subject-relative or culture-relative epistemic relations rather than out-there joints of reality.[18]

This irrealist stance generates a problem. The problem is not, mind you, that the fact that $\sim\mathbb{S}$(because) and $\sim\mathbb{S}(because_p)$ makes them an unsuitable tool to frame the distinction between reality and appearance (which is what we have said about grounding earlier in this chapter). For on my account, it is not the job of explanation to frame this distinction: this job has already been done, in another fashion.

This is rather a methodological problem. For this kind of constitutive explanation might suggest a very naïve methodological stance, viz. that our ordinary and epistemic notion of non-causal explanation interfaces reality from appearances in the sense that by providing instances of such an explanation, we are hereby moving from the less perspicuous to the more perspicuous (if not the absolutely perspicuous); in this sense a constitutive explanation suggests a metaphysically more transparent way to describe what is the case. Ultimately, to provide a chain of explanations is to arrive to what is *really* the case – in the sense that we have articulated in these two chapters.

But here is the kick. As explanation itself is a primarily epistemic phenomenon, we cannot assume that a shift from what is less clear to what is clearer hereby constitutes a shift from the less perspicuous to the more perspicuous. This is, on the contrary, an assumption very hard to argue for: what is enlightening, and what is perspicuous, might be two very different things.

Put in other words, this naïve methodology involves a very crass form of chauvinism. Here is another way in which Sider's case of the blue-red surface is helpful again: as members of our linguistic community, we might find the utterance of (e1) and (e2) enlightening, and thus genuinely explanatory in character; but this is

[18] One may believe my irrealism to be a source of difficulty when characterizing explanation and its formal features; after all, many features of causal explanations are understood vicariously through the study of the formal features of causation itself. And so, that there are no out-there relations for a non-causal explanation to track may make the job harder. I disagree: for even if there was, *pace* me, an out-there grounding aspect of reality, how would we know of its properties? The only access to it, it would seem, is through the association with our ordinary concept of explanation; on this topic, see Maurin (2019).

simply because we are more accustomed to "red" and "blue" predicates as opposed to "bled" and "rue" predicates. The alien tribespeople will find "red" and "blue" mysterious instead, and may thus offer their own "explanations":

e3) s_1 is red because r_1 is bled and r_2 is rue, or
e4) s_2 is blue because b_1 is rue and b_2 is bled

Now, why are these mere "explanations"? Why the scare-quotes? It seems to me that the only non-question-begging reason to have this opinion is that $\mathbb{S}$(is red) and $\mathbb{S}$(is blue), while ~$\mathbb{S}$(is bled) and ~$\mathbb{S}$(is rue). But if, as the naïve methodology suggests, the only reason to hold $\mathbb{S}$(is red) and $\mathbb{S}$(is blue) is because (e1) and (e2) appear to us genuine instances of explanation, then we arrive to a circularity. At this point, the alien tribespeople are more than allowed to reach the contrary conclusion that, because (e3) and (e4) appear to them genuinely explanatory, then $\mathbb{S}$(is bled) and $\mathbb{S}$(is rue), whereas ~$\mathbb{S}$(is red) and ~$\mathbb{S}$(is blue). Debates about the reality of colours then devolve into mutual accusations of obscurity.

In general: if one does not already know how reality is structured (as we, in fact, do not), the alleged existence of individual instances of explanation is not going to prove much. The error of the supporter of the naïve methodology is to read too much into one's personal or cultural biases, whichever they might be (hence why I called it chauvinist). And so, whether red or rue is real (or both, or neither), I reject the idea that the explanations of red in terms of rue and bled, provided by the alien tribespeople, need to be mentioned in scare-quotes: whichever profound metaphysical difference exists between "explanations" and explanations, such difference amounts to *nothing* from a methodological point of view.

Yet this leaves us with the question: how should the metaphysician proceed in their quest to discern reality from appearances using explanations? The answer to this question, or at least an inkling thereof, will conclude the chapter.

3.5.2 Inkling for a Methodology of Metaphysics

There is an important lesson that we need to learn here: explanatory or otherwise bridging principles between classes of sentences cannot perform the job that many philosophers would like them to perform. E.g., a metaphysician of Quinean persuasion might want to "paraphrase away" ontological commitment to a hole by shifting from "there is a hole in my sock" to "the sock is perforated"; but the way I see things, this still does not settle whether it is really the case that there is a hole in my sock. Paraphrase is a lateral move from one truth to another; it is never a decisive move from mere truth to perspicuity; or, more to the point: some paraphrases are, and some are not, but we are in no position to know which ones simply by assessing their material adequacy. In other words: that a paraphrase or an epistemically successful explanation exists, does not decide *per se* whether it is metaphysically illuminating. Another instance of this fact is the one provided in the earlier section about truth: the mere shift from object-language sentences and their metasemantic truth-conditions, of any kind, is not *per se* metaphysically illuminating. A spectacular instance of this phenomenon, for those familiar with the literature on dispositions, is the so-called conditional analysis of dispositions; we'll come back to that.

This doom and gloom is no news, of course. Although I disagree on details, I agree with the general sentiment against what has been occasionally called "linguisticized metaphysics", according to which "deep truths about the universe are to be had by analysing ways we talk about the universe" (Heil, 2012, p. 10). In general, one of the biggest hurdles for contemporary metaphysicians stems from the fact that they find themselves between a rock and a hard place. On the one hand, one must be wary of reading too much metaphysics into one's utterance, whether in everyday, scientific, or philosophical discourse. On the other hand, "I see no way around using language in the theory of ontological commitment" (Schaffer, 2008, p. 9), and the same goes, more generally, in discussing matters of realism and anti-realism in metaphysics.

In my account, this is expressed by the fact that realism and anti-realism concerning a certain discourse cannot simply be decided by the truth of ordinary sentences: more to the point, that some Xs belong to the furniture of reality does not simply amount to "there are Xs" being true; for Xs to be part of the structure of reality, discourse about Xs, whether existential or not, has to be perspicuous. This has a peculiar effect: that whichever bridge-principle (paraphrase, analysis, explanation, etc....) one employs to dismiss X-related discourse cannot demonstrate that Xs are not real.

What then, are we to do? Are metaphysical matters ultimately doomed to be unresolvable? I don't think so. It seems to me that an indirect or "holistic" approach is preferable. Even if there is a culture-relative or even person-relative element that makes individual instances of explanations successful, explanations do not usually occur in a vacuum; rather, the entire explanatory architecture might be evaluated based on its explanatory successes, and the resources needed to achieve them. To the extent in which we might think of theories as bodies of explanations, or explanatory hierarchies, our best shot is to take the sentences at the bottom of our best theories to be metaphysically perspicuous. This suggestion too comes in the spirit of Sider's (2011) extension of Quine's ontological methodology. Because we are rejecting the Ontological Temptation, when discussing matters of realism in metaphysics we are not simply discussing matters of ontological commitment; we are rather discussing matters of structure, and therefore perspicuity. According to Sider (2011, p. 15),

> [t]heories are good insofar as they are simple, explanatorily powerful, integrate with other good theories, and so on. We should believe generally what good theories say; so if a good theory makes an ontological claim, we should believe it. The ontological claim took part in a theoretical success, and therefore inherits a borrowed luster; it merits our belief. This all is familiar; but a believer in structure can say more. A good theory isn't merely likely to be *true*. Its ideology is also likely to carve at the joints. For the conceptual decisions made in adopting that theory – and not just the theory's ontology – were vindicated; those conceptual decisions also took part in a theoretical success, and also inherit a borrowed luster. So we can add to the Quinean advice: regard the ideology of your best theory as carving at the joints.

But we shouldn't stop at ideology, viz. at predicates. Because the $\mathbb{S}$ operator can attach to any subsentential expression, the commitments about reality of all expressions in a candidate sentence can be vindicated, if its credentials are good enough – even pieces of logical machinery like negation and quantification.

This arches back to something I said at the end of the previous chapter, in which I claimed that, despite my resistance to the Ontological Temptation, realism does not come for free; at the time, I did not clarify what this extra effort amounted to. Now I can: rather than trying to settle the reality of Xs by engaging in direct arguments for or against the truth of sentences in the relative discourse (whether existential or not), or by putting forward, in a piece-meal fashion, individual instances of bridging principles such as analysis, paraphrases, or explanations, we must resort to metaphilosophical and holistic evaluations. When a metaphysical theory, understood as a body of explanations, enjoys enough virtues over its rivals, then the sentences to be found at the body of these explanations have the more credentials to be thought as perspicuous. As I said before, when the explanatory questions stop is an interest-relative matter, in metaphysics as anywhere else; but when a theory is subsequently evaluated vis-à-vis its rivals, and there its primitives can be properly evaluated as expressing what is really the case. As in science, this can be a gradual process: the more successes a theory enjoys, the more we should believe in its primitives as expressing genuine aspects of reality's structure. This is very clearly a complicated affair, to be sure, something that can only be undertaken when the candidate theories already exist in a mature enough stage to allow them to be compared. But whoever said that settling matters of realism was easy?

In other words, when we proceed in our explanatory endeavours in metaphysics, we should not conceive ourselves as "following the trail of grounding" like good truffle-dogs; for even if there was a trail to follow (which I doubt), along unionist or separatist lines, we wouldn't have a sense for it (remember Dasgupta's joke about "groundons"). Rather, what we are doing, is tentatively finding better concepts to carve reality into. "Tentatively" is the key word here: this is a holistic and gradual process.

It is also, quite importantly, a defeasible process. For Siderean methodology also inherits this from Quinean metaontology: it is inherently hopeful. For Quine, the hope was that the ontological commitments of sentences in the canonical notation of our best theories indeed correspond to the genuine ontological furniture of the world. For us, the hope is that the primitives of our best theories indeed carve out a genuine aspect of reality. This kind of "Quinean hope", as we might call it, is perhaps motivated by the assumption that reality is to some extent transparent to our theoretical enterprises – that it is neither simpler nor more complex than our more virtuous descriptions of it. What arguments may exist to motivate this assumption, is anyone's guess. But in my opinion, the metaphysician should brace themselves for the possibility that reality, so to speak, might not cooperate. Thus, the aforementioned methodology may provide increasing levels of justifications for believing in a certain structure of reality – whose increases are positively correlated to the successes of the corresponding theory. But absolute certainty, as they say, is not of this world. This is, I believe, a methodological sword of Damocles that the hardcore metaphysician will have to learn to live with.

3.6 Conclusions

I have argued from a metasemantic ground in favour of a factive conception of appearances, according to which, if it appears to be the case that ϕ, then ϕ; whereas it might be the case that ϕ without it really being the case that ϕ – so-called mere appearances. In the previous chapter I have articulated and defended a Siderean non-ontic notion of structure to characterize reality. In this chapter, I have presented a deflationary notion of appearances; for the fact that ϕ to belong to the realm of appearances, all it takes is for "ϕ" to be a Moorean certainty – very simply: a sentence for which there is metasemantic pressure to be taken as true; for some fact ϕ to belong to the realm of mere appearances, all it takes is for "ϕ" to merely true, thus not perspicuous. No shadowy realm of mere appearance is needed, nor any layering aspect of reality bridging the more fundamental to the less fundamental. On the contrary, this deflated notion of appearance and mere appearances is mostly exhausted by the claim that certain sentences are true or merely true, and thus factual.

Amongst others, the metaphysician might employ a family of non-causal explanations to interface reality with appearances. I have introduced at least one: constitutive explanation; but I will introduce more in due time. But readers beware: under this metaphysically deflated framework, constitutive explanation is a primarily epistemic enterprise, and is not *per se* able to detect whether we are moving from the less perspicuous to the more perspicuous. Since I do not believe in any genuine layering of reality, I neither believe that there are objective relations, or more generally aspects of reality, that guide our quest for constitutive explanations; even if there were, there would be no straightforward access to them. In other words, whenever we claim that red and blue explain bled and rue, we do not proceed out of certainty that red and blue are "more fundamental" or "more natural" than their counterparts – those explanations rather function by way of familiarity with certain concepts as opposed to others.

The method I have proposed is inherently holistic and meta-philosophical. As far as the metaphysician is concerned, reality does not constrain our explanations by ruling out individual instances of them; after all, taken in isolation, there's hardly a reason to prefer an explanation from red and blue to bled rather than an explanation from bled and rue to red. But reality still does constrain our explanations, by settling that certain bodies of explanations are inherently more virtuous than others. And so, if we are to believe that some sub-sentential expressions are joint-carving, we should believe that those to be found at the bottom of these particularly virtuous bodies of explanations are. As defeasible as it is, the "Quinean hope" is, now, the hope that the sentences at the bottom of these bodies of explanations are indeed metaphysically perspicuous, and thus reveal the structure of reality. Yet one should keep in mind that this hope is always in principle defeasible.

This inkling for a methodology of metaphysics concludes the theoretical backdrop of the book. With this understanding of the role of the metaphysician and the nature of the world they are supposed to investigate, we can hopefully approach the topic of dispositions from a more favourable angle.

Chapter 4
Dispositional Reality

4.1 Introduction

In the loosest sense of the term, to be a dispositionalist is to take true dispositional ascriptions metaphysically seriously, and to believe that they tell us something about reality. But what exactly? This is where the notion of Dispositional Reality finally enters the picture: with the metaphysical background painted in the previous two chapters, we are now in the position to cast dispositionalism under a new light. In the following three chapters, I will introduce and discuss realism about dispositionality upon this novel background. Let us begin, in this chapter, by introducing Dispositional Reality.

4.2 The Appearance of Dispositionality

4.2.1 A Diverse Family of Dispositions

The realm of appearances is a busy place: it is populated by numbers, times, wrongdoings, trees in people's backyards, ways for things to be, and so much more. It is also populated by dispositions. Ceramic is brittle, in the sense that it is disposed to break under stress. Propane is a flammable gas, in the sense that it is disposed to ignite when exposed to flame. Table salt is soluble, in the sense that it is disposed to dissolve when immersed in water. People have dispositions too: someone may be talkative (viz., disposed to talk), irritable (viz., disposed to anger), or assertive (viz., disposed to impose themselves). Despite what certain philosophers once hoped (Quine, 1973, p. 10), dispositionality is not the exclusive purview of the ordinary or the macroscopic: on the contrary, phenomena of scientific interests frequently receive dispositional characterizations; e.g. according to the Brønsted–Lowry theory of acids and bases, acid compounds are to be understood as proton-

L. Azzano, *Dispositional Reality*, Synthese Library 482,
https://doi.org/10.1007/978-3-031-52625-1_4

donors, viz. disposed to donate protons (H^+ ions), whereas a base is a proton-acceptor, disposed to accept protons; similarly, according to Lewis' theory, an acid is a compound with the peculiar disposition to attach itself to an unshared pair of electrons when in a chemical reaction with another molecule. Or, borrowing another example from physics, electric charge may be characterized as the disposition to attract or repel other charged particles at a certain distance with a force whose magnitude is given by Coulomb's Law[1] – whereas, with much approximation, in quantum electrodynamics electric charge can be understood as a disposition to emit or absorb photons.

I have expressed these dispositions in a very simple way, using locutions such as "disposed to... when..." or "disposition to... if..."; after the first blank comes what we will call a manifestation specification, whereas the second blank should be filled by a stimulus specification. That said, the over-use of this limited terminology may have been unduly reductive, and may in fact have concealed an even wider family of features, arranged in a more sophisticated taxonomy.

It may be worthwhile to explore the multifaceted nature of dispositionality, such as we find it in the realm of appearances. Firstly, one might be dissatisfied with the claim that ceramic is brittle: perhaps it would be more precise to claim that ceramic is *extremely* brittle, in the sense that it takes very little force to break it. This allows for a very common kind of comparative dispositional ascriptions, e.g. when we say that, out of all metals, gold is amongst the most malleable, as it is easy to deform under compressive stress. This is normally taken to imply two things: first, that dispositions come in degrees, and second, that dispositional ascriptions are a highly contextual matter, in the sense that which degree of a disposition warrants an ascription depends on the context of utterance: whether something counts as "brittle" or "malleable" is a highly contextual matter.[2] The notion of a gradable disposition is often cashed out through the idea of determinable/determinate dispositions, in the sense that dispositions are supposed to display some degree of mathematical structure; distinct degrees of a disposition are thus understood as different determinates of the same determinable, with different values plugged in. Whereas in the stimulus and manifestation specifications of determinable dispositions numerical variables may appear, in the specifications of determinate dispositions we will most likely find specifications of quantities as well – up to the maximally specific quantities of the so-called "perfectly determinate" dispositions.

Secondly, the use of the expression "disposed to..." may be limiting in the sense that it excessively focuses on those features that make the subject liable to suffer some kind of change, instead of eliciting it into something else; "solubility" is a

[1] Albeit the most often cited, this is not the only disposition we can associate with electric charge (Kistler, 2020). We will come back to the relation between dispositions and properties such as electric charge in Chap. 6.

[2] Bird (2007a, p. 20) and, more recently, Vetter (2015). Vetter calls "potentialities" the context-insensitive metaphysical backdrop of dispositional ascriptions; but for now, let us focus on the appearance of dispositionality.

ready-made expression to talk about the disposition of, say, sugar to dissolve in certain substances, say, water. But there is no ready-made expression to talk about the disposition of the water to dissolve sugar (but there is a nominalized expression: such a substance is called "solvent"). This doesn't feel right: water and its features are just as important as sugar in a dissolution process; so we may, borrowing from a ultimately Aristotelian distinction, distinguish between active and passive dispositions; active dispositions are occasionally also called "capacities": the capacity of water to dissolve sugar, or the capacity of aspirin to relieve headaches (Cartwright, 1989, p. 3). But whether the distinction cuts any metaphysical ice, remains to be seen; after all, it's not like water survives unscathed the dissolution process: when a sugarcube dissolves in it, a solution is formed.[3]

Thirdly, the example of the sugar and the water suggests that most dispositions do not work alone, but with certain dispositional partners collaborating towards a mutual manifestation, such as the sugarcube's disposition to dissolve in water and the water's disposition to dissolve sugar.[4] With some effort, we may individuate lone dispositions too, which function without partners; e.g. the radioactivity of an unstable nucleus, understood as a disposition to emit radiations and lose energy, would seem to function without the need of other dispositional partners (Mumford & Anjum, 2011, p. 35–36).

Fourthly, we might want to focus on the stimulus and manifestation specifications. While it is perfectly understood that dispositions need to be manifested somehow, one may wonder whether, in some cases, stimuli may not be needed. Some of the previous examples apparently corroborate this line of thought. E.g., a person may be talkative in the sense that they are disposed to talk (this disposition most likely admits of degrees such as the ones discussed above; some people may be highly talkative, while others no to so much; and some people are surely more talkative than others). While it is somewhat clear what is the manifestation of talkativity (viz., talking), it is not so clear what constitutes the stimulus. *Prima facie*, there's no single stimulating factor, but a wide range of circumstances in which a person's talkativity may be sprung into action; and the same goes for irritability (Manley & Wasserman, 2008, p. 72). In other cases, the lack of stimuli may be related to the lack of dispositional partners seen above. As radioactive decay is understood to be a continuous and spontaneous process, the notion of radioactivity as a disposition is not simply that of a disposition without partners, but also without stimulus (Lowe, 2011, p. 22); indeed, one might easily cash out a disposition's stimulus as the coming together of dispositional partners to produce a manifestation,

---

[3]Current orthodoxy is that there is no metaphysically significant distinction between active and passive dispositions; for a dissenting voice, see Marmodoro (2017) and Engelhard (2021).

[4]This notion of reciprocity was first put forward by Martin (1993, 1997) and is by now very popular. It goes by various names in the literature, e.g. amongst others, "Power Net" (Martin, 1997), or "constellation of powers" (Williams, 2019).

so that a disposition without stimulus is, by definition, one without partners.[5] Whether that is the case, there is a further complication: not all stimulus-less, allegedly spontaneous and continuous dispositions are without partners. Here we may want to discuss more closely the notion of electric charge, which I have *prima facie* characterized as the disposition to attract or repel other charged particles at a certain distance with a force whose magnitude is given by Coulomb's Law. But what exactly is the stimulus of this disposition? This is not altogether clear; as in Bird (2007a, p. 22) we might naïvely think of the stimulus as the two particles being at a certain distance, but that would seem to be more like a necessary condition, rather than a causing circumstance for their mutual attraction or repulsion (Lowe, 2011). Furthermore, it's not like a charged particle is standing idly by until another one is put at a certain distance (which distance, by the way? If two particles are concrete in the sense that are always spatiotemporally located, then any two particles are always at a certain distance); perhaps more appropriately, we could conceive electric charge as the continuously and spontaneously manifested disposition to induce an electromagnetic field (Esfeld & Sachse, 2011, p. 38); thus, as a disposition without stimulus. This stimulus-less disposition can operate on its own, or with other dispositional partners, viz. other charged particles, to induce fields with different properties.

Fiftly, and finally, the notion of a disposition or a capacity so understood, may still be unduly restrictive in the sense that it is only one notion in a wider family of features, which includes other kinds of features such as abilities, habituals, tendencies, and the like. There's a kind of family resemblance between these notions: they all display a manifestation specification (perhaps a stimulus specification as well, but not necessarily), and they all involve what is usually called, for lack of a better term, a modal aspect: they do not characterize things as they currently are, but as they may be, or could be in a certain circumstances – they constitute, to use a famous expressions, "threats and promises" about the world (Goodman, 1954, p. 40). Whether these notions constitute various *species* of a more fundamental dispositional *genus*, it is not my place to say.

More generally, I would suggest the reader to resist what is an understandable philosopher's instinct: viz., that of tidying up and simplifying the taxonomy hereby presented – whether by way of eliminating some of these distinctions (viz. the one between active and passive dispositions), selecting one of the two ends as privileged (e.g., the determinate/determinable distinctions), or by introducing a more fundamental dispositional category (e.g., as in the case of Vetter's potentialities).

The reason, simply put, is that we are still dealing about the world of appearances. Given that, as I have characterized in the previous chapter, the realm of appearance is

[5] The first to talk explicitly about dispositional partners was Martin (1993, 1997, 2007), but the idea has now become exceedingly popular; e.g. Heil (2003, 2005) and Bird (2007a); according to Mumford and Anjum (2011, p. 37) "[a] stimulus may often be just a further power that gets added", which indeed seems to suggest that we should think of manifesting powers not as being "stimulated" but as being "partnered"; but also see McKitrick (2013, p. 132), and Vetter's (2015) "joint potentialities".

but the realm of truth and factivity, this claim ultimately amounts to the fact that most, if not all, of the previous dispositional ascriptions are true and factual. The job of the metaphysician, so to speak, comes after. E.g., one may have a metaphysical preference or arguments in favour of determinable dispositions over determinate ones, or *vice versa*; but both determinable and determinate dispositional ascriptions are true. E.g., of a certain particle p with electric charge of c_1, where c_1 is a numerical variable, we may say that p is disposed to exert a force $F = k_e \frac{c_1 c_2}{r^2}$ on another charge c_2 at distance r; yet if charge c_1 assumes a specific numerical value, it is also true that, say, p is disposed to exert a force of 8×10^{-8} N on another charge of 1.6×10^{-19} C at a distance of 5.3×10^{-11} m. Both claims are true. Furthermore, the effort to eliminate certain categories of dispositions is often motivated by the need to pursue a certain philosophical project; e.g., in the debate between Manley and Wasserman (2008, 2011) and Vetter (2013, 2015), it was the need to analyze dispositions through conditionals that made the existence of stimulus-less dispositions problematic: because where are we supposed to find the antecedent for the much-needed conditional, without a stimulus? But we are not at this stage yet: we are merely observing what is the case. And in this context, a *laissez-faire* attitude towards this plurality of dispositions appears to be well-suited.[6]

That said, however, there is a final distinction worth stating, between the more ordinary dispositions and the more theoretical ones. This is not so much of a clear-cut distinction, but a continuum. Despite the existence of a grey area in the middle of this continuum, dispositions to be found at the two extremes are somewhat easy to individuate; e.g., an ordinary disposition might be irritability, water-solubility, or impenetrability, whereas the characterization of electric charge as a capacity to emit or absorb photons, according to the principles of quantum electrodynamics, is probably a good example of a theoretical disposition. These two kinds of dispositional ascriptions are different at least in one important respect: the former are obviously true in a way that the latter are not. Ordinary sentences like "this sugarcube is disposed to dissolve in water" may enjoy the kind of certainty that we have previously characterized as Moorean certainty: viz., the kind of certainty proper of those sentences whose rejection constitutes bad linguistic practice, or a veritable metasemantic error. One may very well argue that the ordinary and pre-theoretical use of dispositional expressions of this very simple variety puts metasemantic pressure against anyone who wishes to reject the correspondent truths. More to the point, one might argue that since solidity can be loosely characterized in a dispositional fashion – viz., as a resistance to the application of external force, or, more aptly, as a disposition to exclude other objects from the object's space – then

[6]This attitude is recently shared in McKitrick (2018, p. 5), who claims that "dispositions are abundant and diverse. [...] [P]hilosophers have been known to make universal generalizations about the nature of dispositions, and, in doing so, limit the domain of dispositional properties. Such generalizations are at odds with the diversity of disposition ascriptions that people tend to make." As this passage makes clear, this diversity within dispositionality for McKitrick clearly transcend considerations of appearance (viz. considerations of truth) and bears a distinctly metaphysical value. We will come back to McKitrick's dispositional pluralism in Chap. 6.

the example of the Moorean certainty, "the wooden plank is solid" discussed in the previous chapter is in fact a case of an ordinary disposition. And the same goes for "this sugarcube is soluble", "aspirin has the capacity to relieve headache", and "gold is malleable". Whichever metaphysical enmity one has towards dispositions, the attribution of these ordinary dispositions is certainly "more credible than any philosopher's argument to the contrary" (Schaffer, 2009, p. 357): if anything is a non-starter, error-theory towards ordinary dispositions surely is.

But that an electrically charged particle has the capacity to absorb or emit photons may not be as obviously true, if anything because its truth requires the truth of quantum electrodynamics (whether *that* is more credible than any philosopher's argument to the contrary, is a serious question about the relation between physics and philosophy that I will not settle here). In general, there is no pre-theoretical and ordinary use of the expressions mentioned in the stimulus and manifestation conditions of such dispositions, that may provide the same degree of certainty as in the case of ordinary dispositions. Of course, once again, there is no sharp cutting-off point between ordinary and theoretical dispositions, if anything because there is no sharp cutting-off point between the "everyday life" in which we observe that sugary things dissolve in water, and the theoretical investigations which lead us to postulate that electrically charged things attract and repel photons.

For our purpose, it does not really matter if the realm of appearance presenting itself before the eyes of the metaphysician has fuzzy boundaries. What matters is that these appearances are often dispositional in character and multifacetedly so. And so, what is the metaphysician to do?

4.2.2 Beyond the Appearance of Dispositionality

The metaphysician may legitimately wonder whether dispositionality is real or mere appearance: in other words, one may formulate a question of realism for dispositions. This is where things get tricky, but after the previous chapters, we have a foundation to step on.

Dispositionality is often expressed in natural languages in a predicative position; several of the aforementioned examples are a testament to that (although not exclusively, e.g. we might say "gold's disposition to deform under compressive stress is a cause of its widespread use"). In a way, the question of realism for dispositionality loosely resembles the question of realism for morality. This fascinating (albeit imperfect) analogy can be exploit for our current purposes. Although never explicitly recognized, it is possible to build a Moorean Argument against dispositional anti-realism which mirrors the Moorean Argument against moral anti-realism. As we have seen, such arguments are notoriously incomplete:

1) Killing babies for fun is wrong
2) ...?
3) Thus, morality is real

4) The sugarcube is disposed to dissolve in water
5) . . .?
6) Thus, dispositionality is real

As the question marks indicate, there is some trouble spelling out premises (2) and (5), to effectively bridge premises to conclusion; Fine's (2001) suggestion was to spell out conclusion (3) as an ontological claim about the existence of moral facts; thus, a way to spell out premise (2) could be: if p, then there is fact that p. Or, we could understand (3) as a claim about the existence of moral properties, and (2) as the introduction of a second-order existential quantifier, ideally ranging over the properties of wrongness.

But such strategies are conceived in sin: namely, the sin of indulging in the Ontological Temptation, according to which the debate about the reality of morality is a debate about the metaphysical status of certain entities, in this case: the existence of moral properties or facts. Yet the moral realists need not embrace (moral) properties or facts, and moral anti-realists need not reject them.

For our purposes, the case of dispositionality in (4)–(6) mirrors the case of morality in (1)–(3). To be sure, I am not claiming that that the cases of morality and dispositions are identical in every respect. Dispositions and moral features are very different beasts, facing different metaphysical problems. The analogy between them is partial, and indeed very abstract. The point I am trying to make here is that, in both cases, our trouble of spelling out the second premise of the Moorean Argument in a way that is acceptable by all anti-realists of the relevant variety, is indicative of a wider phenomenon; viz., that is should be possible to discuss matters of realism towards dispositions, or morality, without it being reduced to a discussion about ontological commitment to a special class of dispositional/moral entities (viz. properties).

After all, if we really wanted to, we *could* conceive the claim that dispositionality is real as the claim that there are dispositional properties or facts, or that dispositional properties or facts are fundamental, or that they enjoy some other privileged metaphysical status. (Incidentally, in the case of dispositionality the focus has been almost exclusively on dispositional properties, as opposed to facts; I suppose a particularly ingenious metaphysician could develop an account of dispositional facts.) Yet, as in the case of morality, we would be wrong in looking at things in this way: just as the moral realist need not embrace moral properties or facts, the realist about dispositionality need not embrace dispositional properties or facts: for at this point, it is unclear why property nominalists, or anyone who rejects the existence of explicitly dispositional entities of any kind, should be denied the right to be a realist about dispositionality. In other words, when discussing the issue of realism for dispositionality, we should not indulge in the Ontological Temptation: the matter of realism for dispositions – as a genuine debate amongst metaphysician – ought not necessarily to be understood as a debate about the ontological commitment to a peculiar class of dispositional entities, properties, facts, or what have you. It may be understood as a debate about the metaphysical value (cashed through the lenses of metaphysical perspicuity) of dispositional ideology.[7]

This is what I am now going to provide: it is finally time to meet Dispositional Reality.

4.3 Meet Dispositional Reality

Consider again sentence (4), the ordinary premise of the Moorean Argument against dispositional anti-realism:

4) The sugarcube is disposed to dissolve in water

We have said that (4) is true, and perhaps even uncontrovertibly true: but is there something else that we can say about it? We can say that it is metaphysically perspicuous, in the sense that it reveals some degree of structure about reality; in fact, as we have seen at length in Chap. 2, this is a convincing way to disarm Moorean Arguments, avoiding error-theory and deflationism. In the case of dispositions, we want to avoid both. So, we may want to frame dispositionalism as the claim that there is a dispositional aspect of reality's structure, which can be revealed in sentences like (4); their truth is not sufficient to settle the matter of realism for dispositions, which is what disarms the argument. A foe of dispositionality, such as a

[7] One could object that the case of dispositions and morality is dissimilar vis-à-vis the Ontological Temptation, and thus the analogy between them not really helpful. To be clear I did concede, in a footnote in Chap. 2, that not all phenomena of metaphysics are on the same boat here, and that the Ontological Temptation may appear more or less dangerous in different contexts. A relevant difference may be epistemic in nature: moral and numerical features of the world belong to a metaphysician's empyreum, in a way that dispositions are not (although, of course, many would disagree). This may in turn generate some differences vis-à-vis the Ontological Temptation. To take an extreme example, consider the debate on the reality of material objects, such as trees. Trees are perceived by humans as discrete and self-contained units in which reality is parsed (e.g. trees are easily countable), in a way that makes the Ontological Temptation less threatening; it doesn't seem very problematic, and almost automatic, to assume that a debate around the reality of trees revolves around the ontic postulation of trees-as-entities. Moral features are different. Whichever our epistemology of morality is, we do not perceive moral features as discrete and self-contained units in which reality is parsed (e.g. "wrongnesses" not so easily countable as trees –as the debate on the supposed repeatability of properties has shown). This makes the Ontological Temptation more pertinent.

Yet even assuming the value of this very broad distinction, it seems to me that dispositions resemble moral features more than they resemble trees. It doesn't seem that we have epistemic access to dispositions as discrete as self-contained units in which reality is carved, and "fragilities" are just as countable as "wrongnesses" (that is to say: problematically). At this at this (admittedly very high) degree of abstraction, the analogy works well enough to drive the same point home: it should be possible to discuss matters of realism towards dispositions or morality, without it being reduced to a discussion about ontological commitment to a special class of dispositional/moral entities.

neo-Humean, can take a sentence such as (4) as a mere truth instead, without being forced to any unwanted conclusion such as (6).

What is this dispositional aspect of reality's structure? Once again, it all starts with the claim that sentence (4) is perspicuous, which, given LS, should be taken to mean to that $\mathbb{S}$(the sugarcube) and $\mathbb{S}$(is disposed to dissolve in water), thus both subsentential components are joint-carving.[8] Remember that we can also use a comparative notion of perspicuity, based on how many of such components are joint-carving; so whether, and to what degree, "the sugarcube is disposed to dissolve in water" is perspicuous, depends not only on the status of "is disposed to dissolve in water", but also that of "the sugarcube". For now, let us simply consider the claim that (4) is perspicuous *tout court*. According to the Siderean non-ontic notion of metaphysical structure that I advocated for, for an expression such as "is disposed to dissolve in water" to be joint-carving, it does not require for it to correspond to any property, neither in a lightweight (e.g. set-theoretical) or in a heavy-weight understanding of the notion; in fact, it does not require that the expression corresponds to any entity at all; the idea that there (4) reveals a dispositional "aspect" or "joint" of reality should not be understood as reifying such notions in any capacity. Rather, to say that reality's structure displays a dispositional "aspect" or "joint" is simply to state that true dispositional ascriptions are underpinned by patterns of similarities which we can express through a metaphysically deflated operator of fact-similarity. E.g., if "sugar is disposed to dissolve in water" is perspicuous, and if, say, "salt is disposed to dissolve in water" is at least true (if not itself perspicuous), then when sugar is disposed to dissolve in water, is like when salt is disposed to dissolve in water. As we have seen at length in Chap. 2, we should resist the temptation to cash out this idea by stating that some entity (what we would normally take to be a "dispositional property") is in fact reoccurring in these cases of similarities.

We can actually say something more. Remember that dispositional ascriptions involve a specification of a partner/stimulus, and a manifestation. And so, the expression "is disposed to dissolve in water" is logically complex; it is unclear exactly how this logical complexity should be expressed; if we simply took (4) to involve a predicate with three argument places, that would entail that dispositional predication is inherently relational, which some people may resist on the ground that dispositions are monadic, as in Martin (1997, p. 216); alternatively we might take "disposed to..." to be some kind of predicate modifier, as Vetter (2015) does for potentialities. This issue of logical form is no matter for the time being: remember that $\mathbb{S}$ attaches to any and all subsentential expressions of a perspicuous sentence, whichever their logical form might be; thus, if (4) is perspicuous, then $\mathbb{S}$(is disposed to), irrespectively of the logic of dispositionality. Suppose, for simplicity, that we cash out such a logical form by positing a three-argument predicate "...is disposed to... if...". Now suppose that the following sentences are true:

[8]Remember that $\mathbb{S}$ can attach to expression of any syntactic category; the above claim only assumes that the grammatical surface of "the sugarcube is disposed to dissolve in water" reflects a logical form of the kind "P(x)".

7) a is disposed to M_1 if S_1
8) b is disposed to M_2 if S_2
9) c is disposed to M_3 if S_3

Depending on whether $\mathbb{S}$(a), $\mathbb{S}$(b), and $\mathbb{S}$(c), (7), (8), and (9) will display various degrees of perspicuity. But now suppose that $\mathbb{S}$(. . .is disposed to. . . if. . .). So, when a is disposed to M_1 if S_1, is like when b is disposed to M_2 if S_2, and also like when c is disposed to M_3 if S_3. What is reoccurring in these facts is not a specific disposition, for the three objects do not share any disposition. What is reoccurring is, so to speak, *dispositionality itself*. This is our first approximation to what I will call Dispositional Reality, viz. the dispositional aspect of reality's structure. As before, Dispositional Reality is not an entity of any kind; rather, to claim that there is Dispositional Reality is to claim that there are patterns of objective similarities underpinning our truthful uses of dispositional ascriptions; in other words, to accept that there is Dispositional Reality is to accept that dispositional predication is metaphysically privileged ideology. As a result, one shouldn't read too much into the existential quantification in claims like "there is Dispositional Reality": this is only for the sake of convenience and expedient communication.

From these admittedly abstract similarities, we can move to more ordinary ones, like the similarities regarding the water-solubility of sugar and salt. E.g., suppose that these two sentences are true.

10) a is disposed to M_1 if S_1
11) b is disposed to M_1 if S_1

Now, it may be the case that $\mathbb{S}$(. . .is disposed to. . . if. . .), but this is still not enough to settle that there are objective similarities regarding any disposition a and b may possess: after all, dispositionality may be metaphysically privileged ideology, but this does not *per se* mean that if two things are equally disposed to be dissolved in Lorenzo's favourite beverage, that constitutes any metaphysically privileged ideology. For that, we need the additional claim that $\mathbb{S}(M_1)$ and $\mathbb{S}(S_1)$ – and so that (10) and (11) are at least equally perspicuous to the extent that $\mathbb{S}$(is disposed to be M_1 if S_1). In that case, we have a similarity to the extent that when a is disposed to M_1 if S_1, is like b is disposed to M_1 if S_1. These patterns of objective similarities are also part of Dispositional Reality.

What about a foe of dispositions? In general, a metaphysician of neo-Humean sympathies will reject this conclusion: they will probably claim that, although (10) and (11) may be both true, they are not underlined by any objective similarity. For a foe of dispositions, how a thing is disposed depends on many different factors, from laws of natures to subjunctive conditionals; perfect duplicates across nomically different worlds may be differently disposed, whereas wildly different things may be similarly disposed. This is no news of course; conceptual progenitors of the LS-version of metaphysical structure that I am employing, e.g. notions of natural or intrinsic properties, may deliver the same result. In the most general formulation: for the friend but not for the foe of dispositionality, dispositional ideology is joint-carving, in the sense that similarly disposed things objectively go together. Just like the anti-realist about modality will not deny that killing babies for fun is wrong (but only that "killing babies for fun is wrong" is perspicuous), similarly the anti-realism towards dispositionality will not deny that the sugarcube is disposed to dissolve in water (but only that "the sugarcube is disposed to dissolve in water" is perspicuous).

As the reader will have picked up, the most crucial highlight of this approach is that it completely eschews the Ontological Temptation; much in the same way the issue of realism for morality does not amount to the existence or fundamentality or essence of moral properties or facts, the question of realism for dispositionality does not amount over the existence or fundamentality or essence of dispositional properties. Call it, if you like, *power metaphysics without powers*. The ramifications, as we will see shorty, are widespread. In the beginning of Chap. 2 I claimed that I do not wish to settle whether reality is constituted by objects, properties, facts, or what have you. We can finally vindicate this claim: my preferred account of metaphysical structure does not involve, but is compatible with, structure being constituted by any such entities. Feel free to plug them in, should the need arise; but it is not an entry fee for entering the club of the friends of powers.

The introduction of Dispositional Reality, so understood, is somewhat original in the literature; but it aims to resolve a tension between ontology and metaphysics that is not of my own making: true dispositional ascriptions may be factual, and may furthermore describe reality in a metaphysically illuminating manner, without hereby corresponding to underlying dispositional properties or facts. This tension traces back to the early days of the literature on dispositions. Here is a peculiar question: why were die-hard nominalists such as Goodman (1954) and Quine (1960) so preoccupied in analyzing away dispositional predication, even when, by their own light, it corresponded not to the ontology, but the ideology of a theory? For example, Goodman's (1954, p. 40) goal is to "explain disposition-terms without any reference to occult powers"; yet he immediately makes it very clear that, strictly speaking, such disposition-terms, as predicates, do not refer to any power, occult or otherwise; in fact for him "the problem about dispositional predicates does not arise from their failing to perform some designatory function that is performed by manifest predicates" (Goodman, 1954, p. 42, fn. 8). The problem for him, as he clarifies in the same page, resides in the fact that such predicates have troubling applicability conditions: dispositional ascriptions appear to be particularly spurious predications.

Similarly, there is a sense in which dispositional ascriptions may be said to be factual, which does not involve the usual ontological baggage carried around by friends of dispositions. I have struggled a little bit in Chap. 3 in trying to develop a clear characterization of factuality; but in short, factuality is the flip side of truth: a sentence "p" is factual, and thus "reports some matter-of-fact" to the extent that "p" is true, and thus that it is the case that p. In this sense, the first preliminary step towards dispositional realism is the endorsement of the fact that dispositional ascriptions are factual in the sense that they express matters-of-fact, as denied in Ryle (1949). (Incidentally, it seems to me that Ryle is in this context using "fact" in a deflated sense, and therefore he is not stating that dispositional ascriptions correspond to heavy-duty facts or properties; but rather that they are factual in a sense loosely similar to mine). Perhaps the only paper to explicitly pick up Ryle's mantle in recent discussion, Heras-Escribano (2017) claims that the newly-dubbed "Rylean dispositionalism does not presuppose any kind of ontology behind the use of a dispositional term", and thus "Rylean dispositionalism is not a metaphysical approach"; the author also calls it "non-factualism". But the way I see it, the lack of dispositional ontology is not sufficient to rule out the metaphysical perspicuity of any true dispositional ascription, and neither it is sufficient to rule out its factuality. In a sense, this approach is falling prey to the Ontological Temptation twice over.

4.4 Innocent vs. Loaded Dispositional Talk

I have claimed that the possibility of Moorean Arguments against dispositional anti-realists has never been explicitly considered in the literature. That said, there is a related concern that has recently emerged amongst friends of dispositions: recent literature has highlighted the importance of the distinction between, for lack of a better expression, *metaphysically innocent* and *metaphysically loaded* dispositional talk.[9] Dispositional Reality can help us frame this distinction more clearly, and with less metaphysical baggage.

In a sense, everyone agrees that there are dispositions – the certain truth of the more ordinary dispositional ascriptions, as above, is proof of that. The metaphysical backdrop may then differ between friends and foes of dispositions – although it is not clear exactly how this backdrop should be formulated. And so, it is not clear how to spell out the "in a sense" in "in a sense, everyone agrees that there are dispositions". Friends of dispositions who are sensible to the distinction between innocent and loaded dispositional talk will probably think of the Moorean Argument (4)–(6) against dispositional anti-realism as particularly insidious and in need of dismantling, as they do not believe that the truth of ordinary dispositional ascriptions is sufficient to settle the matter of the reality of dispositions. (This passage may strike the reader as dialectically contorted: why would a friend of dispositions dismantle an argument against dispositional anti-realist? But look at it this way: a dispositional realist still believes their disagreement with the anti-realist to be a genuine metaphysical disagreement, which cannot be resolved in the way suggested by (4)–(6); in order to establish their position, the friend of dispositions must show that their adversary is something more credible than the quasi-strawman kind of error-theory that would result by claiming that sugar is not soluble.)

Ignoring this distinction leads to an almost-Carnapian dismissive metaontological attitude towards the reality of dispositions as either obvious or pointless. Are there dispositions? Well, according to the widely held Alexander's Dictum, to exist is to have powers – so, yes (assuming that something exists). This is what we may call a *spurious application* of dispositions. According to Bird (2016) such spurious

[9] Crucially, Bird (2016) and Azzano (2019).

applications are dangerously widespread. Here is another example: one may want to study the metaethical principle of "ought-implies-can" through a dispositional lens, but unless the account actively engages with the elements responsible for realism about dispositions, whichever they might be, then the same account will be available to foes of dispositions as well. Lewis-inspired metaphysicians surely believe that things are so-and-so disposed – for them, this simply means that certain counterfactuals are true. This is not to say that this account in metaethics *per se* problematic; but that it does not constitute a genuine application of power metaphysics.

Many dispositionalists have voiced similar concerns throughout the years, myself included; but how exactly to frame this distinction between innocent and loaded dispositional talk, and thus how to understand realism about dispositionality in a way to efficaciously neutralize the Moorean Argument in (4)–(6)? In this context, Bird (2016) introduces the distinction between "predicatory dispositions" and "ontic dispositions" (instance of a more general distinction between "predicatory properties" and "ontic properties). These notions are not rigorously defined, but one easily gets the gist of what Bird means by those terms: predicatory dispositions are ontologically uncommitting entities, close cousins of abundant properties perhaps, such that there's one for (almost) any predicate; whereas ontic dispositions constitute a genuine addition to our ontology. Independently, in (Azzano, 2019) I came up with a similar distinction; there I distinguished between what I called "weak powers", understood as mere products of disquotation – such that, e.g. water has the weak power to dissolve sugar if and only if "water is disposed to dissolve sugar" is true – and "robust powers", viz. a genuinely dispositional property. Crucially, the idea was that weak powers and robust powers were not in a one-to-one correspondence, which would have been sufficient to dismantle the Moorean Argument in (4)–(6) under the assumption that realism for powers was about the existence of robust powers.

The predicatory/ontic and the weak/robust divide are two instances of a more widespread distinction amongst friends of dispositions, albeit one not too easy to pinpoint exactly; this distinction is usually framed as the distinction between mere "dispositions" on the one hand, and genuine "dispositional properties", or "powers", or "potencies", or "potentialities" on the other; besides the papers already mentioned, also see Borghini and Williams (2008, p. 23–24), Williams (2019, p. 46–48), Contessa (2019, p. 7–8), and Friend and Kimpton-Nye (2023). Putting forward this distinction is a conceited effort in meta-metaphysics, viz. a way to recapture the distinction between innocent and loaded dispositional talk, and thus express how friends of dispositionality can put forward their claim in a non-trivial manner. Firstly: the dispositions, as I will call them, simply correspond to what I have called the appearance of dispositionality: that things possess dispositions, in the sense, ultimately amounts to the almost trivial claim that many dispositional are true: thus the admission of dispositions is but the rejection of error-theory. Considered in themselves, dispositions are usually understood as non-committing quasi-entities, which merely result from taking dispositional ascriptions as being the case; they are, again using an expression from Azzano (2019), the mere product of disquotation. Indeed Contessa (2019, p. 7) explicitly protests that "dispositions are not entities". Then there are powers, as I will call them, which are presumed to be metaphysically privileged properties, in some sense or another, whose admission makes for genuine realism about dispositionality. Although their metaphysical privilege varies from account to account, whatever makes a power a power is what must be deployed for an application of dispositionality to be non-spurious – to make it unavailable to the foes of dispositions (viz. those who believe that there are no powers).

When trying to cash out the distinction between dispositions and powers, and thus trying to understand the privilege that powers enjoy over dispositions, we encounter a metaphysical landscape very similar to the one we surveyed to distinguish between appearance and reality in Chaps. 2 and 3. Traditionally, the most popular claim is the essentialist one, according to which powers, or genuinely dispositional properties, have a dispositional essence – viz., the so-called dispositional essentialism.[10] Grounding-based accounts of powers have been gaining popularity in recent times, according to which powers ground dispositions.[11] Alternatively, truthmaking has been put forward, so that powers are understood as truthmakers, of modalizations/conditionals, or dispositional ascriptions directly.[12] Finally, some other times the special ingredient of a metaphysics of powers is more vaguely defined as powers being "natural", "sparse", "fundamental", or "basic" properties – or a combination thereof–, without further elaboration.

As diverse as this family of solutions is, most of them suffer, in some degree or another, from a crucial flaw. They indulge too much in the Ontological Temptation: in this case, to think about the reality of dispositionality as an issue about the metaphysical status of a special class of entities (which we have dubbed "powers") – be it an issue of essence, grounding, or aptness for truthmaking. The very fact that we are trying to approach the issue in this way, by separating innocent and loaded metaphysical talk through the distinction between "mere dispositions" and "genuine powers", is the root of the difficulty. As realism about morality is unduly restricted by being framed as realism about moral entities, so is realism about dispositionality being restricted by being framed as realism about dispositional properties. Case in point: Bird (2016, p. 10) explicitly claims that a nominalist is "someone who says that there are no ontic properties", and thus no ontic dispositions and no powers. But a nominalist should be able to claim that powers are real too.

---

[10] "Dispositional essentialism" was originally conceived as a kind of essentialism about natural kinds, as opposed to properties, e.g. Ellis (2001) and Molnar (2003), but it has since then been popularized as a position about the dispositional essence of properties, as in Bird (2007a); position also received extensive discussion in Yates (2013, 2018a). So understood, dispositional essentialism is a close cousin of the "causal structuralist" view of properties, according to which properties essentially have causal and nomic roles, and can therefore be individuated by their position in a suitable second-order structure, as in Shoemaker (1980), Swoyer (1982), and Hawthorne (2001).

[11] As far as I know the first instance of this view can be found in Shoemaker (1980) who claims that "a thing's powers or dispositions are distinct from, because 'grounded in,' its intrinsic properties". But the position has only been more recently developed: with some differences amongst them, see Tugby (2012, 2021, 2022), Coates (2020, 2022), and Kimpton-Nye (2021). Talk of dispositional properties (or powers) "bestowing" dispositions perhaps falls in this category as well, e.g. Heil (2003, p. 79).

[12] Jacobs (2011), Contessa (2019).

As always, some accounts are more problematic than others. A profoundly problematic position is the one to be found in Tugby (2021), according to which properties ground dispositions.[13] Unlike neo-Humean metaphysics, in which the dispositions supervene, or are grounded upon, the total distribution of properties across space-time, on this account an individual property instance is all that it takes to ground a disposition instance (Tugby, 2021, p. 20); in fact, this kind of grounding sufficiency may very well be the special ingredient separating power metaphysics from its neo-Humean adversary, as argued in Azzano (2019, 2020). Tugby's problems emerge with the introduction of a particular kind of grounding; according to Tugby (2021, p. 7), grounding is a relation between entities that, given the standard characterization, allows the grounded entities to be genuine existents, numerically different from what is doing the grounding. This move reproduces all the flaws of a layered conception of reality, as criticized in Chaps. 2 and 3; in particular, it reproduces all the uncertainties regarding the metaphysical status of the "shadowy" derivative existents: after all, with the idea of dispositions being grounded derivative entities, we may want to recapture the notion of dispositions as metaphysically non-committing quasi-entities; yet they still are some entity or another in the layering of reality, up to the point of Tugby (2021, p. 8–9) calling them "real [and] irreducible" – a claim whose metaphysical significance is dubious, given that, on his approach, the mental properties of the Physicalist also are "real and irreducible". This is an old problem, of course, one independent from Tugby's position: the culprit is the claim that matters of realism are to be settled by the positioning of entities along a fundamentality hierarchy. In this respect, Tugby (2021) falls for an Ontological Temptation twice, by reifying both powers *and* dispositions.[14] And with the explicit goal in mind to avoid such a troublesome reification, some friends of powers prefer to deploy truthmaking instead of grounding, e.g. Jacobs (2011), Contessa (2019), but also see Azzano (2019). Of course, that still requires truthmaking entities.

Such difficulties can, in principle, be avoided without giving up much of this conceptual machinery, at least not completely. As we have seen, there are metaphysically deflated operationalist conceptions of grounding and truthmaking, that – although not entirely harmless – can assuage many of the worries generated by layering. As far as I know, they have not been so far employed for this purposes. As for dispositional essentialism, it is in principle possible to switch from an objectual to an irreducibly generic notion of essence, perhaps along the lines of Correia (2006), to generate a more lightweight metaphysics of dispositionality: such a position has recently been developed and advocated in Vogt (2022).

[13]Tugby's (2021) terminological choices are different from mines. What I call "dispositions" he calls "powers"; the properties doing the grounding, although presumably "ontic" in Bird's sense, are not themselves dispositional but qualitative –presumably to avoid a grounding circularity as it appears in certain variants of dispositional essentialism.

[14]Tugby (2021) does however claim that his framework allows for a nominalist take on dispositionality. I am dubious of the results however; I will come back to this shortly. Furthermore, in Chap. 6 (Sect. 6.3.2), I will discuss some crucial differences between my account and Tugby's.

Given Dispositional Reality, another road is available, one which may allow us to recast the distinction between innocent and loaded dispositional talk in clearer terms, and may be able to deflate the admittedly problematic distinction between powers/dispositions, or ontic/predicatory dispositions, or robust/weak powers, as they might be called. In fact, these two goals are achieved at the same time by noticing that we shouldn't understand the difference between innocent and loaded dispositional talk as two distinct portions of dispositional discourse, corresponding to two different classes of dispositional entities: the distinction ultimately amounts to different conversational attitudes towards dispositional discourse itself.

As we have seen in Chap. 2, shifting interests make the implications of our utterances unavoidably context-relative; in some contexts, we may utter "p" to merely entail that "p" is true, or that it is the case that p; while in others, we may utter "p" to also imply that "p" is perspicuous, and thus revealing of the structure of reality. Such shifting interests may have been at the source of the feeling of miscommunication that we might encounter in certain Moorean Arguments; such shifting interests are also responsible for the distinction between metaphysically innocent and loaded dispositional talk. With "innocent dispositional talk" I mean dispositional discourse which is uttered in a context which merely implies the truth of what is being said; on the contrary, with "loaded dispositional talk" I mean dispositional discourse which is uttered in a context which also implies that the relevant sentences are metaphysically perspicuous – and thus reveal the presence of Dispositional Reality.

One may wonder how exactly we are to decide when the use of dispositional talk in the philosophy room is innocent or loaded. There may be some back-and-forth here, but it seems to me that there often are enough context-cues to make a reasonable decision. E.g., someone putting forward the Alexander's Dictum as a criterion for existence in a discussion otherwise unrelated to the issue of realism for powers, is probably engaging in metaphysically innocent dispositional talk, and simply providing a true biconditional bridging matters of existence with matters of dispositionality. Sometimes a miscommunication amongst philosophers may make things clearer. E.g.:

Philosopher (1): "to exist is to have powers"
Philosopher (2): "no, I don't think that powers exist"

In this case, the misgivings of Philosopher (2) seem to be misplaced. But in other philosophical contexts, specifically ones in which the issue of realism for powers is on the table, dispositional talk appears to have a deeper metaphysical significance; and in such contexts, the protestation of Philosopher (2) does not appear to be misplaced. For example:

Philosopher (1): "the properties of subatomic particles are powers"
Philosopher (2): "no, I don't think that powers exist"

I hope, but I cannot have the certainty, that these two exchanges strike the reader the same way they strike me: the former, but not the latter, being a miscommunication. At any rate, I will not provide a list of such contexts, nor I will generalize too much about them. I think we have a good enough ability to spot spurious deployments of dispositionality, so we can proceed on a case-to-case basis. But notice this: under these lenses, the deployment of a ready-made lexicographical distinction between "powers" and "dispositions" can help the informed reader making the right decision; by using the term "power", one may signal that they are speaking with the intention of speaking perspicuously; and by using "disposition", one may signal that they are speaking without bothering about perspicuity. (Keep in mind that my account of Dispositional Reality is meant to be revolutionary, as opposed to hermeneutical: I am not claiming that friends of dispositions actually frame their position as one revolving around perspicuity and structure. They patently do not, so I am not claiming that friends of dispositionality in fact use "powers" to signal their intention to speak perspicuously; but they do sometimes use it to signal their intention to speak in a "metaphysically loaded way" about dispositionality, which is cashed out, on my account, through perspicuity). I will not exploit the lexicographical distinction between "power" and "disposition" in what follows; in fact, this very distinction between the "genuine" powers and the "quasi-entities" dispositions can be abandoned; to engage in innocent dispositional talk is not to talk about "innocent entities" (viz. non-fundamental, derivative entities): the contrary opinion is the result of the quasi-correspondence theory of truth I criticized in Chaps. 2 and 3 (what Rayo has called Metaphysicalism), an all too naïve picture of the interface between language and reality. When talking innocently about disposition, one is merely saying something true (or attempting to), all the while leaving metaphysical matters about the reality of dispositions at the door. There's a further twist, however. For what I have said about disposition also holds for powers, viz. the hypothetical objects of metaphysically loaded dispositional talk: to engage in metaphysically loaded dispositional talk is not to talk about a privileged metaphysical category of so-called powers, the "philosopher's dispositions" (Williams, 2019, p. 47)[15]; to engage in loaded dispositional talk ultimately amounts to engaging in dispositional talk, all the while entailing, if not explicitly declaring, the perspicuity of the relative sentences.

Ultimately, the distinction between metaphysically innocent vs. loaded dispositional talk is based on the "sparsist" perception that "satisfaction of a predicate is not a sufficient condition for the existence of a *real* property distinguished by that predicate" (Ellis & Lierse, 1994, p. 9, my emphasis on "real").[16] On my account this is epitomized by the distinction between truth and perspicuity, such that true predication need not carve or directly correspond to any portion of reality. Since Lewis (1983), we usually deploy heavy-duty properties to draw this distinction between the "real" properties and the mere properties-as-sets required for the truth of the corresponding sentences. But Dispositional Reality is a way to draw this distinction, and thus to deploy the distinction between innocent vs. loaded dispositional talk, without such a reification: in Chap. 2, we have put under scrutiny the very

[15] The expression is allegedly borrowed from Prior (1985, p. 1).

[16] For an explicit endorsement of "sparsism", see Williams (2019, p. 18). For a dissenting opinion, McKitrick (2018); I will discuss McKitrick's stance in Chap. 6.

idea that in a perspicuous predicate-ascription, the predicate corresponds to an entity. The consequences are those mentioned above: my account does not envision the abundant quasi-entities occasionally called "dispositions", but neither does it envision the "real" and "genuine" entities occasionally called "powers".

This allows us to introduce one of the most crucial features of our account of Dispositional Reality: that of a power metaphysics without powers.

4.5 Power Metaphysics Without Powers

One of the main claims discussed in this book is that (some fragment of) dispositional talk is perspicuous and thus revealing of a dispositional backbone of reality – what we have called Dispositional Reality. Properly understood, the perspicuity of dispositional talk does not require the reification of dispositions as properties or entities of any other kind: as such, our account of Dispositional Reality is a metaphysical position that does not *per se* have an ontological component. This has widespread consequences on many areas of power metaphysics, and metaphysics in general, which we now turn to discuss.

4.5.1 Dispositionality and Nominalism

Realism concerning dispositionality, when looked under the lenses of Dispositional Reality, completely eschews the Ontological Temptation, in the sense that it does not require discussing the existence, fundamentality, or essence or a "dispositional property", nor any other entity posited for that purpose: in fact, neither the truth nor perspicuity of dispositional ascriptions requires such entities. Therefore, although my account is not *per se* a form of nominalism (viz. of anti-realism about properties), it is surely compatible with such a position. As it turns out, the combination between the two has a certain appeal, and I am somewhat sympathetic to it.

It is worth noticing that Dispositional Reality is not the first attempt to achieve this result; previous attempts can be found in Whittle (2009), Vetter (2020), and Vogt (2022); Tugby (2021) also discusses a nominalist-friendly variant of his grounding theory of powers, inspired, as some others in this list, by the nominalist strain to be originally found in Shoemaker (1980) – a strain subsequently disavowed by its very author, in Shoemaker (1998, p. 64). That said, the vast majority of friends of powers are non-nominalists, and postulate some kind of property; some of them are trope theorists, others believe in universals, and some of them are even Platonists; perhaps the most lightweight conception of properties endorsed by a friend of dispositions is McKitrick's (2018, p. 52) "agnostic disjunctivism" about properties, according to which "universals, sets of tropes, sets of objects, or perhaps some other entities, play the *property* role, and any of these disjuncts might be playing that role in any particular case".[17]

For a while, the lack of nominalist options may have led the friends of dispositionality to think that powers and nominalism was an unholy marriage; and hopeful statements such as the one in Mellor (1974, p. 157) "I shall put my case in terms of dispositions being 'real' properties. Nominalists will have to rephrase it [. . .] Nominalism is not the issue here", were usually thought to be either too naïve or too programmatic to have any philosophical weight. Yet, all things considered, no real argument existed to settle things one way or the other.

It is true, in my opinion, that some combinations of dispositionalism and nominalism are not very inviting. Yet, as I will argue, others are; so the alleged compatibility between dispositionalism and nominalism crucially depends on details. For example, somewhat ironically some variants of dispositional *cum* nominalism err by way of ontological profligacy. That would be the case of Matthew Tugby, according to which "the powers of things ground the properties of those things" (Tugby, 2021, p. 21). Remember that grounding, for Tugby, is supposed to be a relation between entities, and, most importantly, a relation between entities of the same "ontological category", such that "a thing's instantiating a property and a thing's possessing a power are both *facts* or *states of affairs* of a certain sort" (Tugby, 2021, p. 11); and so, independently from any further metaphysical detail, and independently from my worries concerning grounding and layering, one exactly wonders whether a nominalist would be satisfied by such an inconspicuous transition from "properties" to "powers".

Other attempts threaten to throw away the baby with the bathwater; e.g. there seems to be something problematic in Whittle's (2009) suggestion that more than fundamental than dispositional ascriptions are subjunctive conditionals, which is perhaps more in line with a metaphysics of primitive subjunctive facts, as in Lange (2009), rather than one of powers. In my approach, this worry translates into the question: if dispositional-talk is so easily dispensable, what reason is there to think that it is perspicuous? This is not a way to get rid of powers-as-properties; it is a way to get rid of dispositional predication *tout court.*[18]

---

[17] In the debate on dispositionality, traditional trope theorists are Martin (1997), Heil and Martin (1999), and Heil (2003, 2012) –although Heil refuses to employ the term "trope", which he considers intertwined with DC William's problematic one-category metaphysics of particulars-as-bundles of tropes. Molnar (2003) is explicitly a trope theorist; while more recent attempts to classify powers as tropes can be found in Hakkarainen and Keinänen (2021) and Coates (2022). That said, many of friends of dispositionality happily conceive powers as universals; most paradigmatically, Ellis (2001), Mumford (2004); some of them even conceive them as Platonic (viz. possibly uninstantiated) universals, as Bird (2007a) and Tugby (2013) –who explicitly argues in favour of this option. Finally, according to Lowe's (2005) four-category ontology, there are both tropes *and* universals.

[18] I will come back to these matters, including the metaphysical value of linking dispositions with conditionals, in the next chapter.

Related to this point, here is a completely wrong way to conceive a nominalizing strategy for the dispositionalist (I am not saying that anyone endorsed this strategy, but one may get the feeling that those sensible to Shoemaker's nominalist strain may be, at least to some extent, drawn to it). One may think that a dispositionalist might get rid of properties by cashing out attributions of ordinary or scientific properties in dispositional terms; e.g. if there was a way to bridge ascriptions of electric charge with certain dispositional ascriptions (more on that in the next chapters), then that would be a good way to nominalize away the property of electric charge. But this is a pipedream: to simply explain, or ground, or otherwise analyze ascriptions of electric charge as ascriptions of dispositions does nothing to settle the issue of nominalism: for if electric charge and dispositions are expressed in a loosely similar fashion (e.g., through predication), then whatever reason which makes dispositional ascriptions compatible with nominalism, will also make ascriptions of charge equally compatible with nominalism. Nothing is achieved here. Simply shifting from non-dispositional to dispositional predication (or *vice versa*) does not settle anything concerning the reality of properties (nor the reality of dispositionality, for that matter).

As I said, however, I have some sympathy for the nominalist option. But the kind of nominalism I have in mind is of a different sort entirely. To understand the point, assume that the following three sentences are true.

12) a has the property of being disposed to M if S
13) a instantiates the disposition to M if S
14) a is disposed to M if S

Because, as we have seen, the $\mathbb{S}$ operator can attach to expressions of any syntactic category, all three sentences may reveal the presence of Dispositional Reality; e.g. we can ask whether $\mathbb{S}$(being disposed to M if S), whether $\mathbb{S}$(the disposition to M if S), and whether $\mathbb{S}$(is disposed to M if S). But in some of these sentences there are *other* pieces of conceptual machinery which may be of interests to the metaphysician, and which do not appear in others: in (12) there's talk of properties, while in (13) there's talk of instantiation. In (14) there's neither: (14) is a simple predicative truth.

The kind of nominalism most naturally expressible by my account of Dispositional Reality claims that (12) and (13) are less perspicuous than (14); more generally, it is the position according to which property-talk (and related conceptual machinery, such as instantiation) is non-perspicuous, and thus does not tell us anything about the structure of reality. I think it qualifies as a version of the position sometimes disparagingly called "Ostrich Nominalism"; name-calling aside, Ostrich Nominalism is a fiercely attractive position; loosely put, it is the position according to which simple predicative truths are fundamental. Thus, it would probably submit the following explanations:

15) a has the property of being disposed to M if S because a is disposed to M if S
16) a instantiates the disposition to M if S because a is disposed to M if S

The case of dispositions is but an instance of a more general pattern of nominalistic explanations:

17) the ball has the property of being red because the ball is red
18) the ball instantiates redness because the ball is red

As the reader might remember from Chap. 3, I do not believe that the presence of individual instances of explanation like (15)–(18) may settle much by themselves. To see whether they constitute genuine explanations, and thus to verify the appeal of this version of nominalism, one needs to take a step back and observe the whole explanatory infrastructure. This is where the virtues of (15)–(18) become apparent: explanatory principles such as these are extremely powerful, because moving left-to-right, potentially obscure metaphysical terminology relating to propertyhood and instantiation simply disappears: and so all correspondent issues are deflated. When it comes to the evaluation of theories, understood as bodies of explanations, such a simplifying move is sure to score some points. (Of course the issue is still open as to whether $\mathbb{S}$(is red); but whether redness is real and whether propertyhood is real are two very distinct questions.) One may protest against this account on the ground that Tarskian semantics at least commits us to properties-as-sets (this is, I take it, the only kernel of truth behind Armstrong's famous point that properties are needed to provide semantic values to predicate): yet, as submitted in Chap. 3, there's little to no reason to think that truth-condition biconditionals move us from the less perspicuous to the more perspicuous, or that Tarskian semantic meta-language is perspicuous at all.

But maybe I assume too much. I do not expect the reader to be converted to this combination of Dispositional Reality and Ostrich Nominalism, especially on such a scant basis. That said, I hope to have shown that this a coherent and perhaps even inviting strategy to pursue. In what follows, I will remain somewhat agnostic on the metaphysical status of propertyhood, instantiation, and other beasts of metaphysics. Another thing that has been established, I hope, is that the relevance of these matters is perhaps less decisive than originally thought: neither truth nor perspicuity of predicative sentences, dispositional or otherwise, is reason enough to believe in properties, instantiation, or any other similar aspect of reality; neither they require reification of any such power or property as a constituent entity of reality.[19]

[19] That said, my conclusion has an important corollary for those who would like to discern that much more detail in reality's structure: the price of admission for genuine powers-as-entities, or universals themselves, is now that much higher. This is not surprising: after all, the troubles encountered by the supporters of the One Over Many arguments show how difficult it is to argue in favour of universals simply by pointing out the need to explain predicative or similarity facts. Similarly, that the world around us is teeming with dispositions is not enough to convince that we ought to crystallize this dispositionality in a separate class of powers-as-entities.

4.5.2 *Shedding Some Weight*

Whether or not we frame my account of Dispositional Reality as a positively nominalistic position, my ontologically deflated approach has a peculiar feature: it allows to pursue the dispositionalist project as a primarily explanatory enterprise, all the while shedding some metaphysical weight that is currently burdening it. In Chap. 3 I have articulated the notion of metaphysical theories as bodies of explanation (and the methodology for evaluating them); and the explanatory aspect of my account of Dispositional Reality will come back as a protagonist in the next chapters. For now, I would like to focus on the metaphysical weight being shed.

With this I mean to say that, within the context of Dispositional Reality, many discussions regarding the metaphysical status of powers-as-entities can be partially or entirely deflated, and some of them can indeed be safely ignored. This is not to say that such debates are pointless *tout court*, and that it was foolish to engage with them in the first place: within the context of a certain conceptions of dispositionality, they make sense (and I myself engaged with them at various stages); but for the supporter of Dispositional Reality, many of them do not have the purported importance they seemed to have before; that said, some of such problems indeed do persist in my account, albeit in a somewhat mutated form.

4.5.2.1 The Metaphysical Status of Powers as Properties

This is quite straightforward. Even without the explicitly nominalistic component, my account does not deal in power-as-properties; unless one endeavours to explicitly rephrase dispositional-talk in order to mention properties (e.g. "gold has the property of being disposed to deform under compressive stress"), then any account primarily dealing with the perspicuity of dispositional ascriptions (e.g. "gold is disposed to deform under compressive stress") need not worry about that. The nominalistic component suggested before is only needed to understand property-talk whenever it is explicitly added. Yet I do not wish it force it on my readers: someone skeptical of Ostrich Nominalism, or nominalism in general, could then substitute that account with one more to their liking (for example, a property-realist could flip the explanatory principle (15) on its head, and give a different account of propertyhood and instantation). The same, *caeteris paribus*, holds for instantiation. In other words: realism about dispositionality requires admitting the joint-carving nature of dispositional terms, but which other terms are joint-carving in a true sentence – and thus what kind of dispositional talk is more perspicuous, is another matter entirely (given that $\mathbb{S}$ works on any syntactic category means that we do not have to worry that in the two sides of the explanatory principle in (15), dispositional terms perform two different syntactic roles).

In conclusion, the issue of realism of dispositionality and one's account of properties and instantiation are orthogonal issues; if there are alleged intersections between them, they will be evaluated on a case-to-case basis.

4.5.2.2 Directedness

Perhaps the most well-known intersection between dispositionalist metaphysics and theory of properties in the literature is to be found in Tugby (2013), which advances Platonism (understood in the Armstrongian sense according to which properties can exist uninstantiated) as the preferable candidate for a theory of powers-as-properties. Tugby's argument functions, amongst other things, by taking into consideration the so-called directedness of dispositional properties.

This is where we encounter one of the most cryptic and elusive aspects of dispositionalist metaphysics. Powers are sometimes said to be "directed" or "oriented" towards their manifestations, as in Mumford (1999), and Molnar (2003); Tugby (2013) also calls it "orientation"; Martin (1997), perhaps one of the first to point out this peculiar feature of dispositionality, called it "directedness" and "selectedness", but offered very little metaphysical elaboration for a notion that might have been, in his intention, a primitive affair. Various efforts were made to articulate the notion of directedness through something more familiar, but without much success; dispositional directedness was linked to the intentionality of mental state, suggesting that the physical world, to the extent in which it is dispositional, also displays an intentional or quasi-intentional aspect, as in Place (1996) and Molnar (2003); alternatively, an analogy between powers and vectors was deployed, ultimately based on the consideration that "[v]ectors have a direction [. . .] This is helpful because a power will have a direction - that towards which it is disposed" ((Mumford & Anjum, 2011, p. 24). Neither option is particularly appealing.[20] Thus, when friends of dispositionality spell out dispositional directedness as a relation, it is often implicit that they are thinking about a *sui generis* dispositional relation – mind you, not one relating the disposed object, for the power is a monadic property *per se*, but some kind of second-order relation between the power, stimulus/partners, and the manifestation.

Besides generic worries of unintelligibility, there is a specific problem plaguing the notion of directedness *qua* relation: assuming that powers can be latent, viz. they can be instantiated without being manifested, and assuming that directedness relate a power's manifestation (amongst other things), many instances of dispositional directedness will have a missing *relatum*; this is usually called a "Meinongian problem", as it vaguely implies that the manifestation of a latent power is but a shadowy non-existent. The most common reply to this problem is the so-called type-level reply; according to it, manifestations should not be understood as causal goings-ons or occurrences, such as events or processes, but types of occurrences, repeatable entities of some kind or another. Powers are thus directed towards type of manifestations, instead of specific tokens of manifestations, and the former are not

[20] Against an intentional-like understanding of dispositional directedness, see Mumford (1999). Against the analogy between the directedness of powers and the direction of vectors, see Massin (2009, p. 572–574), Bird (2016), and Pechlivanidi and Psillos (n.d.).

existentially dependent on any of the latter.[21] That said, however, the existence of a type may be dependent on the existence of at least one token, which may thus re-introduce the Meinongian problem in the case of a thing disposed to M, under the assumption that nothing is M; this is where, allegedly, Tugby's Platonism comes to the rescue.

I am not going to judge the level of danger posed by the Meinongian problem, nor the viability of the type-level reply, nor, for that matter, Tugby's Platonism. I only wish to point out the level of ontological sophistication required to concede that there is a problem in the first place: directedness must be understood as a genuine relation, and powers and manifestations must be reified up to the point of being *relata*.[22] I do not reify the existence of powers and manifestations, so I surely am not going to reify the relation between the two. The account I propose of Dispositional Reality has no way to clearly reconstruct this difficulty – if anything because taking dispositional ascriptions as perspicuous has no bearing on the reality of a phenomenon that we may call "dispositional directedness"; if anything, I only accept the claim that dispositions are so-and-so directed in the following sense:

19) a has a disposition directed towards M because a is disposed to M

The theoretical virtues of (19), again, consist in the fact that in move from left to right, potentially mysterious items of metaphysics simply disappear.

That said, there is a residual difficulty for Dispositional Reality. I reject the often-hidden assumption that the logical complexity of dispositional predication ought to be parsed into the kind of (pseudo-)relational ontology submitted by those who talk about "dispositional directionality"; but such a logical complexity is not metaphysically insignificant, either: whether "disposed to..." is cashed out as a relational predicate or a predicate modifier, it still requires term for the manifestation to be completed. That could generate a problem. Consider the following toy-world; there are four objects: a, b, c, and d. As it happens, a, b, and c are M, whereas d is not M; however, d is disposed to M. Assuming that $\mathbb{S}$(M), it will be the case that when a is M is like when b is M, which is also like when c is M; however, "M" also appears in "d is disposed to M", possibly in the same syntactical role, e.g. if "disposed to..." is understood as a predicate modifier; coupled with the fact that our account of perspicuity is not particularly sensitive to matters of logical form, it will follow that when a is M is like when b is M, which is also like when c is M... which is also like when d is disposed to M. And yet, by postulation, d is not M. Slightly different would be the case which originally solicited a Platonic reaction: suppose that a is disposed to M, but nothing is M; in that case, there will be no fact of the form "x is M" for us to build our fact-similarity with the fact that a is disposed to M. But still,

[21] Ellis (2001, p. 132–135), Bird (2007a, p. 104–108), and Handfield (2008). Additionally, Mumford (2004, p. 194–195) offers partially independent motivation for the type-level reply.

[22] It is therefore no surprise that enemies of universals usually reject directedness as a relation; most famously, according to Martin (1997, p. 216) "dispositionality is *not* a relation between what is dispositional and what is its manifestation."

suppose that, although nothing is M, both a and b are disposed to M; so, if 𝕊(M), it will still be the case that when a is disposed to M, it is like when b is disposed to M – there still is an M-aspect of reality, although nothing is M.

Two things are worth pointing out; first, this problem is not like the old Meinongian problem: it does not threaten us with the presence of any non-existent manifestation or transcendent universal; in fact, it does not threaten with any entity at all. Secondly, it is not even clear whether, and to which extent, this is a problem at all. In the first case, that when a is M is like when d is disposed to M, does not entail, *contra* our initial stipulation, that d is M; this fact-similarity only entails that (loosely speaking) the perspicuity of "a is disposed to M" also reveals the reality of M, and that in "a is M" and "d is disposed to M", this M-aspect of reality is reoccurring. The consequences of the second case are also somewhat hard to evaluate: it only amounts to the fact that the 𝕊(M) may manifest itself in the presence of certain patterns of similarities even if nothing is M. This phenomenon is not exclusive of dispositionality, mind you. Suppose that nothing is M, although it is indeed the case that possibly a is M, and possibly b is M; if you accept that 𝕊(M), it would then follow that when possibly a is M, is like when possibly b is M; thus, there is a modal pattern of M-hood, even though nothing is M. Whether, and to what extent, this residual issue is problematic, remains to be seen; but it is *prima facie* less pressing than the original Meinongian difficulty on the traditional understanding of directedness as a relation.[23]

[23] Here is an interesting consequence. Substitutionally opaque contexts are cases in which, from the joint-carving nature of certain expressions, we cannot infer the presence of worldly patterns. Suppose that I am realist about God, in the sense that 𝕊(God); now consider the following true sentences:

a) Pope Benedict XIII is called "the servant of God"
b) The Virgin Mary is called "the mother of God"

given the joint-carving status of "God", we may want to infer that when Pope Benedict XIII is called "the servant of God" is like when the Virgin Mary is called "the mother of God". But would that be wise? It seems to me that the only God-related similarity between the two facts, is that these two people have partially similar names –something that does not require any belief in God to be established. Loosely put, sentences (a) and (b) are not about God, and even if perspicuous, do not reveal that God is part of reality's structure; cases like (a) and (b) involve what are traditionally called cases of referential opacity, in the sense that "God" does not appear in a referential position: the kind of "metaphysical opacity" I mean here is a categorical extension of that notion –based on the idea, established in Chap. 2, that it is not the role of expressions, of any category, in a perspicuous sentence, to refer or correspond to certain entities. If an expression α appears in a metaphysically opaque context in a sentence p; even if it appears in another sentence q, and even if it is the case that 𝕊(α), from that we cannot infer that when p, is like when q. Whether, and to what extent, expressions within the scope of modal or dispositional operators are used opaquely in this sense, remains to be seen; assuming that they *are* indeed used in an opaque environment, the residual problem would disappear. I leave this matter for another time –but I will briefly come back to "metaphysical opacity" at the end of Chap. 8, although for a different purpose.

4.5.2.3 The Essence and Individuation of Powers

To further complicate discussions of directedness was the fact that the directedness of powers is often considered an essential, or otherwise constitutive or essential features, as in Armstrong (1996, p. 69) and Molnar (2003, p. 60). As soon as dispositional properties entered the ontological arena, discussions erupted about their nature, identity and individuation conditions; closely related to the claim that powers are essentially directed to their manifestations, is the claim that powers get their identity and individuation conditions from their stimulus and manifestation specifications: powers, we are told, are essentially powers *for* certain behaviours in certain circumstances. Similar claims are emphasized in dispositional essentialism, as in Bird (2007b), or the causal structuralist view: many of such positions explicitly claim that dispositional properties get their identity relationally, through their position in a second-order structure.[24]

Coupled with a monist assumption, viz. the assumption that all properties are dispositional, this feature of powers generates a veritable metaphysical minefield, where accusations of impredicativity, vicious circularities, and infinite regresses run rampant. There's no single argument here, but a loosely connected family thereof. First of all, Swinburne (1980) suggests an epistemic regress argument based on the idea that, since we access powers through their manifestations, if all properties are powers "the epistemic buck is continually passed" (Tugby, 2014, p. 2) without any knowledge of properties ever being achieved. Most famously, there is also the so-called "always packing, never travelling" objection, an actuality regress argument from Armstrong (1996, p. 80; 2005, p. 314), according to which if all properties were powers, the world would lack in actuality – a worry easily resolved simply by noticing, as in Mumford (2004, p. 174) that powers are not possibilities: "to be potent [. . .] is to be actual". That said, the most discussed argument in this family is Lowe's (2005) identity regress argument, according to which, if power get their identity from their manifestations, but all manifestations are in turn powers, no power would ever get its identity fixed "because each property owes its identity to another". As the literature was quick to point out[25] there is an ambiguity in the use of the term "identity" here, which crucially alters the validity of this complaint and of Bird's (2007b) reply to it. The problem could be that powers have impredicative identity conditions, or that powers are only circularly individuated, or, even more abstractly, that there is a circularity in how powers have their nature fixed relationally through their "directedness". To this "nature regress argument" (Ingthorsson, 2015) are often associated even more abstract worries; e.g. it is occasionally submitted that a world only populated by powers would be a world lacking in reality or being.[26] As for metaphysical individuation, I have recently argued, in Azzano

[24] But also see Mumford (2004, p. 171).

[25] Lowe (2010), Jacobs (2011), Ingthorsson (2015) and Busse (2021).

[26] E.g., Campbell (1976, p. 93), Robinson (1982, p. 114–115), Heil (2003, p. 98/108). But also see Molnar (2003, p. 176).

(2023), that once the notion is properly understood, Lowe's regress argument can be solved in the way suggested by Bird (2007b).

These are difficult and multifaceted issues, which cannot be entirely resolved in this context. But no worries: once again, it is not my job to convince the reader one way or the other; I would only like to point out how these difficulties are reliant on a certain understanding of dispositionality, viz. one with powers-as-properties, which can be bestowed relational or structural objectual essences, and can enter relations of metaphysical individuation, understood as a worldly relation between entities – paradigmatically, as in Lowe (2010, p. 9). It is very dubious whether any of these difficulties will persist, once my account of Dispositional Reality is adopted: clearly, without powers-as-entities, there's nothing to ask identity conditions for; as for the notion of metaphysical individuation, which underpins how a certain entity can be individuated, it makes little sense to search for a deflated operationalist understanding of it, as we can do with metaphysical grounding: what are we individuating, if not entities? As for essence or nature regresses, there is no explicitly essentialist component in my account of Dispositional Reality so far.[27] We could of course follow Vogt (2022) in her shift from objectual essence to generic essences; but it is far from clear whether any regress can be reproduced in this new environment.

4.5.2.4 Dispositional and Categorical Properties

Traditionally, there is a flip side to dispositional properties: categorical properties. If dispositional properties are curious beasts, categorical properties are even more so: back at a time when the instantiation of a dispositional property was understood as the satisfaction of a certain conditional, and thus when dispositional properties were understood as characterizing their bearer as it could have been, or would have been under certain counterfactual circumstances, categorical properties were understood as being unconditional actual qualifications of their bearers. As such, dispositional and categorical properties were conceived as mutually exclusive categories. Based on the idea that modal or conditional properties had to be grounded in non-modal unconditional properties, it was usually assumed that dispositional properties needed "categorical bases" to be metaphysically legitimate.[28]

This set-up was first explicitly questioned in the well-known opening passage in Mellor (1974). A realist about dispositional properties will find the need of categorical bases for dispositions question-begging against their position: after all, to say that something is soluble is not to say how it would or could be. It is to say something actual and unconditional about it: that it is soluble. In other words, "dispositions are, in a very obvious way, categorical" (Mumford, 1998, p. 64). As we have seen, this is

[27] I will more extensively argue against the inclusion of such essentialist components in Chap. 8.

[28] The notion of a categorical property or base was first introduced in Armstrong (1968, 1969); although, by his own admission (Armstrong, 2005, p. 319), he borrowed an expression originally found in Price (1953, p. 321–322).

the dispositionalist's go-to strategy to dispel Armstrong's "always packing, never travelling" objection against dispositions. Relatedly, it was argued in in Ellis and Lierse (1994, p. 30–31) that even if some dispositions indeed need to be grounded (as is presumably the case of ordinary macro-dispositions such as fragility or solubility), there is little reason to think, beyond a generalized prejudice against dispositions, that a special class of non-dispositional "categorical" properties will be needed to do the job.

Without the need of categorical basis, one may believe that the need of categorical properties also disappeared; yet the divide between dispositional and categorical properties persisted, although it became increasingly nebulous and difficult to articulate. Sometimes the distinction between the two kinds of properties is put forward with the help of paradigmatic examples, yet this kind of strategy is severely flawed, even as a mere heuristic tool, for the simple fact that intuitions about such paradigmatic examples (e.g. about which property is dispositional, and which is categorical) wildly fluctuate even amongst philosophers; thus, I will not provide such examples. One thing is sure: we cannot operate the dispositional/categorical divide simply by observing predicates: even assuming that we have, based on our investigation on the appearance of dispositionality at the beginning of this chapter, a clear understanding of what is, and what is not, dispositional predication, such a distinction would cut no metaphysical ice. As this chapter has made exceedingly clear, there is no presumption amongst friends of dispositionality that dispositional predicates and dispositional properties are in a one-to-one correspondence; so it is more than likely that some dispositional property may be hiding behind a non-dispositional predicate, and that some non-dispositional property may be hiding behind a dispositional predicate. We have already seen efforts to characterize the dispositionality of properties, especially in their "ontic", or "robust" variety; for every such characterization, we were often offered a correspondent characterization of a categorical property as well. Thus, a tripartite taxonomy of positions immediately presents itself: one the one hand "Dispositional Monism", according to which there only are dispositional properties; "Categorical Monism", according to which there only are categorical properties, or "Dualism" or the "Mixed View", according to which there are both. Arguments for and against these positions abound (for example: the identity regresses discussed above may be understood as arguing in favour of Dualism, and again Dispositional Monism).

Where does that leave us, with respect to Dispositional Reality? First of all, just as the present account eliminates any need to characterize any inherently dispositional entity, it also eliminates any need to characterize any inherently categorical entity. But this is not to say that the matter of the dispositional/categorical divide is entirely discarded. There is a residual issue here, which may be expressed as such: a supporter of Dispositional Reality is not automatically forced to endorse the claim that every true dispositional ascription reveals Dispositional Reality: for some dispositional ascriptions may not be perspicuous, or at least, not absolutely perspicuous; perhaps this is the case of macro-dispositions, or extrinsic dispositions – or some other kind of spurious dispositions. The methodology for metaphysics put forward in Chap. 3 suggests the following way forward: we can discover something about the structure of reality by exploring the ideological and conceptual choices at the bottom of the explanatory hierarchy of a particularly virtuous body of explanations. Thus, a supporter of Dispositional Reality who also espouses Dispositional Monism may be understood as believing that all non-dispositional ascriptions can ultimately and advantageously be understood in dispositional terms. A Categorical Monist believes the contrary (and thus denies that any dispositional ascription is perspicuous – it denies that there is Dispositional Reality in the first place); while a Dualist rejects both claims: there is Dispositional Reality, but there is a predicative aspect of reality which is not perspicuously expressed through dispositional ideology.

This is an issue concerning the extent of Dispositional Reality, which will be discussed in Chap. 6.

4.5.2.5 Powerful Qualities, and the Identity Theory

Uncertainties about how to characterize the dispositional/categorical divide are so widespread that, while the two categories were usually through to be mutually exclusive, there are those who believe that properties are both dispositional and categorical (or, to complicate things even further, "qualitative").[29] This is the so-called Powerful Qualities view (perhaps the most widespread variant of the Mixed View, according to which there are both dispositional and categorical property, albeit not separately). Traditionally the Powerful Qualities View was expressed as an Identity Theory; in this context, the identity does not merely amount to the fact that every dispositional property is identical to a categorical one, but that for each property, its dispositionality is identical to its categoricity. Of an arbitrary property P, Heil (2003, p. 111) has to say:

> P is simultaneously dispositional and qualitative; P's dispositionality and qualitativity are not aspects or properties of P; P's dispositionality, P_d, is P's qualitativity, P_q, and each of these is P: $P_d = P_q = P$.

How to understand this extremely robust identity without collapsing this position on a metaphysically more neutral view is the subject of ongoing debate.[30] Furthermore, given the trouble that we have encountered understanding the notion of "dispositionality" and "categoricity" it is not even clear what a property will look like which is both.[31] Suffices to say that the Identity Theory is, together with the topic of directedness, one of the more elusive aspect of dispositionalist metaphysics, a profoundly complicated affair whose theoretical pay-off is very much unclear.

[29] Martin (1993, 1997, 2007), Heil and Martin (1999), Heil (2003), Strawson (2008), Schroer (2010, 2013), Jacobs (2011), Ingthorsson (2013), Taylor (2013, 2018), Williams (2019), Giannotti (2019).

[30] For the most recent attempt, Giannotti (2019).

[31] For a surprisingly anticlimactic conclusion, see Taylor (2018); further developments can be found in Giannotti (2019, 2021), and Azzano (2019).

The adoption of my approach based on Dispositional Reality has a simplifying effect. It does not reify properties, and it surely does not reify their dispositionalities or categoricities – so the entire problem concerning their identification cannot get off the ground in the first place. Yet there is a looser sense in which an identity between the dispositional and the categorical may be expressible within the confines of my framework, which may be developed by deploying, rather than standard objectual identity, an operator of generalized identity, as in Correia (2010, 2016) and Dorr (2016); this, perhaps, may take the form of Rayo's (2013) "just-is" operator – as further explored in Linnebo (2014) – which, again, as a sentential operator is flanked by two sentences and does not require the reification of what is being identified. How does generalized identity fit within my metaphysical set-up is a fascinating question that cannot be extensively discussed here; yet we are in a position to say something about it. E.g., borrowing Heil's (2003) own example, we may say:

Id) the ball being spherical *just is* the ball being disposed to roll

Many features of the Powerful Qualities View, *qua* Identity Theory, are preserved here. E.g., being generalized identity undoubtably symmetric (Rayo, 2014, p. 518), no priority is stated between the qualitative and the dispositional side, as desired (Martin, 1997, p. 216). Furthermore, given that no ordinary objectual identity is employed in (Id), there are no two things which are identified – no two properties, or properties of properties, or aspects, or sides, or what have you; which is what a supporter of Dispositional Reality desires: yet (Id) very clearly expresses an identification between the qualitative and the dispositional.

The symmetry of generalized identity has a more general consequence: it does not enrich reality with a problematic layering between the more fundamental and the less fundamental, nor it provides, *per se*, any kind of metaphysically underpinning for a non-causal explanation under the banner of explanatory realism (see Chap. 3). Given (Id), it may still be a matter of contention whether one side explains the other, and which one (a supporter of the Powerful Qualities View would probably deny either direction of explanation). Indeed, principles like (Id) may hold in the case of some explanations like the ones put forward in the next chapters. This goes to show that under this approach, the claim that the categorical is identical to the dispositional may very well turn out to be something less problematic than originally claimed, although, perhaps, also something less contentious and interesting. Whether supporters of the Powerful Qualities View would be satisfied by expressing their position along the lines of (Id) remains to be seen – as far as I can tell, this particular kind of identification between the dispositional and the qualitative has never been discussed in the literature.[32]

[32] I will discuss (Id) again in Chap. 6.

4.6 Conclusions

This is enough, for our first encounter with Dispositional Reality. Let us recall how we got here: in previous chapters I have submitted that reality may not an amorphous lump, but displays certain repeating patterns, in other words certain aspects, while in this chapter I have submitted that one of these aspects may indeed be dispositionality – Dispositional Reality. To believe that there is Dispositional Reality is to believe that (at least some) dispositional talk is metaphysically perspicuous – which is more than saying that it is true, something which may be safely claimed by foes of dispositions as well. That a dispositional ascription is perspicuous entails that a repetition of the same dispositional expression is underpinned by genuine fact-similarities (with a lightweight conception of facts in mind: let us not put too much emphasis on the notion of an "aspect" or "joint" of reality). Thus, this is an ontologically inconspicuous way to reconstruct the distinction between metaphysically innocent and metaphysically loaded dispositional talk.

This dispositional aspect of reality, quite importantly, is not a thing, and thus not a property of any kind. Paired with the metaphysical set-up previously exposed, there are many ways for a friend of Dispositional Reality to be a nominalist; I have a sympathy for a variant of Ostrich Nominalism, understood as the position according to which simple predicative truths (dispositional or not) are fundamental. Another virtue of this account is that it sharply deflates many problematic issues that have plagued dispositionalist metaphysicians in recent years (directedness, identity regresses, identity theory, etc...), or reframes them in a much more manageable fashion. This goes to show that at least some of these issues are not deep and foundational issues of power metaphysics, but rather the result of a faulty perspective; more specifically, the result of an unduly reification of Dispositional Reality. As such, they might be the result of indulging too much in the Ontological Temptation. For once you have introduced dispositional entities in your metaphysics, presumably in the form of properties, questions start to pop up: what kind of properties are they? How do they receive their identity? Are they somehow "directed" to their manifestations? How? What makes them dispositional? Are there non-dispositional properties as well? What is the relation between them? Not so for the supporter of Dispositional Reality, for which such questions either have no answer, or very simple answers.

In the following two chapters, I will explore additional features of the account. In Chap. 5, I will approach the decisive question: what reasons do we have to think that there is Dispositional Reality after all? Even accepting the meta-metaphysical set-up so far provided, one still must motivate the claim that all, or some, dispositional talk is metaphysically perspicuous; in doing so, we will need to pick up the metaphysical methodology put forward in Chap. 3. This is where the inherently explanatory element of my version of dispositionalism will come to the forefront.

In Chap. 6 I will evaluate the extent of Dispositional Reality: what kind of dispositional talk – within the variety hereby presented – is to be taken as perspicuous? Relatedly, what other kind of non-overtly dispositional predication is ultimately underpinned by Dispositional Reality?

Chapter 5
In Favour of Dispositional Reality

5.1 Introduction

Why should we believe in Dispositional Reality? This chapter is dedicated to the arguments for dispositionalism, as they have been presented in the literature, and reframed by the supporter of Dispositional Reality.

On the face of it, the history of the arguments in favour of power metaphysics is a confusing one: there is no "master argument" to rehabilitate powers in contemporary literature, and considerations in their favour come from a variety of debates in metaphysics and philosophy of science, with little to no indication that such considerations in fact point in the same direction; furthermore, some of the more popular arguments, like the Argument from Science have been proven to be highly contentious. The situation is in fact so methodologically uncertain that some dispositionalists now simply prefer to present their own position, and let its own virtues speak for itself; perhaps this is motivated by a broadly abductivist methodology in metaphysics, yet sometimes the inference is so timid to be almost non-existent. This is epitomized in Williams' (2019, p. 4) "soft sell" attitude, according to which

> there is a coherent alternative to neo-Humeanism. I want to tell you about the alternative, have you think about it, and have you come to see the neo-Humean metaphysic as replaceable. And I want you to cast a critical eye over both. That way your choice is genuine. And though at times I indicate aspects of the neo-Humean metaphysic that look more like flaws than features, it is not my plan to construct devastating objections against it. In fact, we should be sceptical of anything presented as a devastating objection against neo-Humeanism as a whole: it is a solid, coherent, and all-round attractive metaphysic.[1]

[1] A similarly non-confrontational attitude, although perhaps not to this extent, can be found in Vetter (2015), in which she presents her own dispositional modal semantics while withdrawing from any critical comparison with its alternatives, most notably standard possible-world modal semantics.

L. Azzano, *Dispositional Reality*, Synthese Library 482,
https://doi.org/10.1007/978-3-031-52625-1_5

My methodological framework, first sketched in Chap. 3 and now applied to the case of dispositions, is meant to illustrate an operating procedure for the dispositionalist to justify their position, and potentially sway doubters.

This is not to say, mind you, that in this chapter I will provide an exhaustive argument in favour of dispositionalism; I will rather illustrate how such argument ought to be formulated. The plan for the chapter is therefore fairly simple. In Sect. 5.2 I will illustrate my methodology – centred around what I will call an Argument from Indispensability. Simply put, this is an elaboration of the vaguely neo-Quinean idea that the theoretical virtues of dispositionalism, as an explanatory hypothesis, are meant to provide partial corroboration for the view itself. In Sect. 5.3 I will discuss how such a methodology reflects on some of the previous considerations in favour of dispositions. Finally, in Sect. 5.4 I will offer some concluding remarks and a guideline for the following chapters.

5.2 Indispensability and Explanatory Dispositionalism

It goes without saying that which argument in favour of dispositions one will find convincing, greatly depends on one's approach to matters of realism or antirealism, either in general or in the specific case of dispositions. In recent literature, arguments meant to convince about the reality of dispositions take multiple forms: some of them are meant to convince that there are properties which are essentially dispositional, while others are meant to convince that dispositional properties are fundamental or irreducible; other lines of argument are more "ontological" in character, arguing that dispositions deserve to be admitted in one's inventory of the world, perhaps by virtue of a cost-benefit analysis not dissimilar from Lewis' own assessment concerning possible worlds. These lines of argument are not always clearly distinguished, and the boundaries between them are often fuzzy.

What about the friend of Dispositional Reality? How are they meant to convince their audience? In this section, I am going to present a type of argument in favour of Dispositional Reality, which I will refer to with the term Argument from Indispensability, and then discuss some of its most salient methodological consequences. As any other argument, the Argument from Indispensability does not function without a theoretical backdrop; in this case, this is the meta-metaphysical framework presented in Chaps. 2 and 3, of which the notion of Dispositional Reality presented in Chap. 4 is an instance. I would like to point out that, strictly speaking, the Argument from Indispensability is not an argument but a class of arguments (I would have called it an argument-scheme, but the lack of a proper formal regimentation makes me wary of that term); I will not detail nor endorse any specific argument within the class, so strictly speaking no argument is presented in favour of Dispositional Reality. I am more interested in articulating what kind of reasoning is best suited for a realist about dispositions with a sympathy for Dispositional Reality: in other words, I want to present the kind of argument that could reasonably motivate the endorsement of Dispositional Reality.

5.2.1 The Argument from Indispensability

To believe in Dispositional Reality is to believe, first and foremost, that at least some dispositional locutions are joint-carving, and therefore that some fragment of dispositional talk is metaphysically perspicuous. Yet, even if one was willing to discuss dispositionality in these terms, why should they believe that some fragment of dispositional talk is indeed perspicuous? The presence of extensionally adequate bridge-principles between dispositional ascriptions and something else is not by itself sufficient to get the job done (this is famously the case of the so-called "conditional analysis" of dispositions – which I will discuss later); after all, a biconditional can be read both ways, and its material adequacy by itself does not allow us to decide which direction – if any – constitutes an increase in perspicuity. *Prima facie* we do not know whether dispositionality is part of reality's structure, as opposed to conditionals, laws of nature, or perhaps possible worlds, so there's hardly a reason to impose a direction upon the correspondent bridge-principles as metaphysically privileged: to do so would be, patently, to put the cart before the horse.

The way I see it, metaphysical theories about dispositionality and related notions should be understood as body of explanations. Bridge-principles such as the conditional analysis are to be understood thusly as explanatory principles. Justification for a certain theory, as opposed to its rivals, comes from the holistic evaluation of such a theory vis-à-vis such rivals, an evaluation based on its explanatory successes and the resources required to achieve such successes: when a metaphysical theory, so understood, enjoys enough theoretical virtues over its rivals, then its explanatory primitives have the best credentials to be thought as perspicuous – and thus as expressing what is really the case. This can be understood as a categorical extension of Quine's methodology for ontology, according to which our best shot at perspicuity is to take the sentences at the bottom of the most virtuous theories to be metaphysically perspicuous, and thus their components to correctly carve the structure of reality. The expression "correctly" requires a caveat: this is an ongoing and always-in-principle-defeasible project in which increasing levels of justifications for believing in a certain portion of reality's structure are positively correlated to the explanatory successes of a certain theory vis-à-vis its rivals. In other words: the more successes a theory enjoys, the more we should believe in its primitives as expressing genuine aspects of reality's structure.

With this in mind, belief in Dispositional Reality would be justified were we to find dispositional locutions at the bottom of the explanatory hierarchy of our best theory. The supporter of Dispositional Reality could thus try to argue that dispositional locutions are indispensable, at least in the very broad sense that they cannot be replaced without a significant loss of some theoretical virtues (e.g. if that theoretical virtue is explanatory power, then we will be dealing with explanatory indispensability; otherwise, some different kind of indispensability will be discussed): doing away with dispositionality, in short, leaves you with a worse theory. This sketch of a methodology surely leaves much to be desired. Perhaps the most urgent question to be answered here is the following: to the extent that the theories under consideration for the evaluation of Dispositional Reality are to be considered as body of explanations, what are the *explananda*? What kind of explanatory enterprise is being pursued here? And these theories I mentioned, what are they theories of?

For our purpose, there is no single phenomena to be explained by metaphysical theories developed by friends and foes of dispositionality (to be clear, the "non-Humeans" and their "neo-Humean" adversaries), but a family thereof. Such *explananda* belong to the category of what we have called appearances, viz. facts in the deflated sense of what is the case, or what can be expressed by true declarative sentences. Here is a list of such appearances that some reader might find familiar.

Regularity The world, at least at the macro-level that we experience in our everyday life, and as investigated by many scientific disciplines, is neither random nor chaotic: on the contrary, it appears to display a certain degree of regularity – although it might not appear as such if one does not know where to look. If our approach to metaphysical structure suggests that it is the job of the metaphysician to find patterns in nature, then this is something that they have in common with the scientist: the world of appearance is in fact teeming with patterns; some of these patterns are man-made (e.g., cars stopping at red lights), and thus are metaphysically superficial phenomena more pertinent to the study of the normative and the intentional; but other patterns of regularity (e.g., sugar dissolving when immersed in water) are, so to speak, built into nature itself, and are thus much more interesting for our current purposes.

One could then ask the legitimate question of *why* exactly these patterns occur. To be sure, several explanatory endeavours are hiding behind this simple question: what currently interests us, with some approximation, is not merely the question as to why the same pattern of behaviour is repeatedly exhibited in multiple cases under similar conditions. E.g., we may be asking why sugary things are disposed to dissolve when immersed in water – a question that cannot be understood as a request for a causal explanation of any specific event.[2]

Nomicity Although the world displays certain patterns of regularity, we know that not all regularities are alike: some of them are merely accidental, while others are not – they are usually said to be "lawful" or "nomic" because, unlike accidental regularities, they are underpinned by a law of nature. There are laws of all kinds: from the most general and famous principles in physics, to those of the various special sciences, and finally down to specific and still tentative principles.[3] Furthermore, as in Friend (2016, p. 136) laws of nature can go by different names, so that it may slightly unclear what exactly counts as a law of nature. But no matter: the notion of a law of nature is crucial in philosophy of science and certain parts of metaphysics, although it is very far from clear what kind of thing a law of nature is supposed to be – if a thing at all. More importantly, that there is a certain law is what explains the

[2]For the many peculiarities of a supposed non-causal explanation of regularities, see Chap. 7.

[3]The literature on laws is so massive that it is impossible to even briefly encapsulate; see Carroll (2021) for examples and an up-to-date summary of the debate.

regularities in a certain pattern of behaviour, and thus it is what makes it non-accidental. Although different philosophical sensibilities might frame this kind of explanation in different terms, there is some degree of confidence that this kind of explanation is non-causal in character: "[o]n the one hand, there are good reasons to think that laws explain their instances. On the other hand, there are good reasons to think that laws do not cause their instances" (Emery, 2019, p. 1535). We might tentatively express nomicity through a sentential law-making operator "L:", to be read as "it is a law that…"; so if $\forall x(Px \rightarrow Qx)$ is a non-accidental regularity, it is the case that $L{:}\forall x(Px \rightarrow Qx)$.[4] We are still within the realm of factuality and truth, and thus the realm of appearance: to the extent that there are true law-statements, nomicity itself is an appearance to be accounted for.

Modality There is a modal strength to non-accidental nomic regularities, and thus to laws of nature, which may be expressed and recaptured in a variety of ways. Most famously, only nomic regularities must pass what we may call a "counterfactual test", in the sense that only to nomic regularities and laws of nature correspond true subjunctive conditionals. Those too are appearances to be accounted for. Of course, subjunctive conditionals are accompanied by other related modalities, such as "physical" or "natural" possibilities and necessities, and they too will require some explaining.

Dispositionality Finally, as we have seen at length, things are so and so disposed. The world of appearance is teeming with dispositionality.

It is difficult to succinctly summarize what it is that ties together these four families of notions; they are sometimes called the "nomic family" of features of the world of appearance. Some of these features may strike as elusive, and it surely doesn't help that they have often been explained one in terms of the other. The interplay between these notions is notoriously complicated, and cannot be exhaustively mapped in these pages. E.g., the correlation between non-accidental regularities and laws of nature is profoundly entrenched in the literature, as is their mutual correlation with counterfactual conditionals (viz., subjunctive conditionals with a false antecedent).[5] It is widely held that law somehow explain their instances, and thus some non-accidental regularities[6] – although on the other hand, for certain philosophers laws themselves are to be accounted for in terms of regularities: this might generate what has occasionally been perceived as an explanatory circularity.[7]

---

[4]This regimentation is a *prima facie* choice that corresponds more transparently to the natural language expression "it is a law that…", but it doesn't preclude the use of other, potentially more perspicuous, devices to express nomicity. More on this topic in Chap. 8.

[5]For an introduction on these topics, see Van Fraassen (1989), chapters 1 and 2, and Carroll (2021).

[6]For an explicit defence of this claim, see Emery (2019, p. 1541–1544).

[7]For this explanatory circularity concerning neo-Humean regularist account of laws, especially Lewis' Best System Account, see Armstrong (1983, p. 102), Maudlin (2007, p. 172), and Lange (2013, p. 256); for a popular reply that distinguishes between two kinds of explanation to dismantle the circularity, see Loewer (2012).

What is more, laws of nature notoriously enrich reality with their own special brand of necessity; sometimes nomicity has been accounted as first and foremost a kind of necessitation[8]; alternatively, laws and their necessity have been explained in terms of primitive subjunctive facts.[9] Contrariwise, there has been dispositional account of laws of nature[10], subjunctive conditionals[11] and modalities of any kind[12], and *vice versa.*[13] And this doesn't even begin to take into account the mutual relations between these notions and other members of the nomic family that I have not considered so far, such as causation and objective chance. In other words, it's a tangle of explanatory endeavours in which one's *explanans* is another's *explanandum*, although one must always keep in mind that these accounts are hardly presented in a unified metaphysical framework. In dealing with this sticky situation, I am reminded of a passage from Lewis (1983, p. 197–198), originally meant to address an entirely different topic, which I think fits quite nicely:

> [w]e have a tight little family of interdefinables. What to do? (1) We could Quine the lot, give over the entire family as unintelligible and dispensable. That would be absurd. (2) We could take one or another member of the family, it scarcely matters which, as primitive. That would be quite acceptable, I think, but disappointing. Or (3) we could somehow break in from outside. That would be best.

Let us list our options. Option (1) indeed strikes me as absurd: to completely ditch the nomic family as unintelligible just because it is proving to be difficult to sort seems like an overreaction. Unlike Lewis, I will not devote much time to Option (3); not because I take it to be an unfeasible strategy, but because I am not aware of any outside player that could feasibly rearrange the nomic family in a new fashion; naturally, this only speaks of the current state of the debate (and, perhaps, of a lack of imagination on my part), but so be it. Whether or not is the best option, as Lewis claims, it is not forthcoming for the time being.

---

[8]This occurs in the so-called DTA theory of laws of nature: Dretske (1977), Tooley (1977), and Armstrong (1978, 1983).

[9]Primarily, Lange (2009).

[10]Cartwright (1989, 1999), Hüttemann (1998), Lipton (1999), Kistler (2003), and Bird (2007a).

[11]Jacobs (2010). Like Lange (2009), Jacobs wishes to turn the traditional relation between subjunctive conditionals and sentential modalities like necessity on its head, using a strategy first deployed in Williamson (2004, 2007).

[12]Mondadori and Morton (1976), Mumford (2004, p. 168–170), Borghini and Williams (2008), Contessa (2010), and Vetter (2015).

[13]E.g., Whittle (2009) and Manley's (2012) "modalism" discusses the possibility that, metaphysically, dispositionality exhausts itself with subjunctive conditionals. But, as we will later in this chapter, the correlation between dispositions and conditionals is historically centrepiece in discussions concerning dispositionality and its reality. The most spectacular example of rearranging the nomic family by downgrading dispositions is to be found in Goodman (1947) and Chisholm (1946), where an attempt to explain dispositions in terms of counterfactuals conditionals quickly turns into an attempt to explain counterfactual conditionals in terms of laws of nature and nomic regularities.

We are therefore left with Option (2); viz. pick a primitive and try to get everything else from that starting point. Such efforts already exist, of course: in the literature there are reconstructions of the nomic family starting from primitive laws of nature[14], primitive subjunctive facts[15], primitive possibilities and/or necessities[16], and primitive dispositionality.[17] Others like to assume primitive distributions of qualities of qualities across space-time, thus primitive regularities, and start from there to get everything else.[18] Options abound. In the above passage Lewis claimed that Option (2) could be "disappointing"; he doesn't expand as to why that is the case, but I suppose that the reason might be the following: if indeed the members of a certain family are interdefinable, there's an air of arbitrariness to the choice of primitive.

Such a manifest lack of motivation for our choice in theoretical primitives is an uncomfortable situation for a metaphysician.

Luckily for us, this is not really the case of the nomic family, and this is where my methodology provides a clear way forward. For the members of the nomic family are not as easily interdefinable as the passage may suggest. Their organizations of the nomic family do not mirror one another, in the sense that they do not display the same explanatory structure: indeed, all of the above strategies have their own peculiar strengths and weaknesses, virtues to exploit and weak spots to neutralize. And so, by properly assessing the features of these different approaches, the choice of primitive in a nomic family is not as arbitrary as one could think. Indeed, we can leverage strengths and weaknesses of one's metaphysical theory to pick the best one, and thus (tentatively) crown its primitive as real.

Let us see how this strategy might function in practice. Let us consider two theories, so understood as explanatory hierarchies, one of them called Lewisian Metaphysics (LM), and the other one called Power Metaphysics (PM). As the name suggests, LM is an ordinary kind of neo-Humean metaphysics: it starts with a certain distribution of qualities across space-time, which gives rise to certain regularities. Ultimately, there is no explanation as to why some fundamental regularities obtain as opposed to others (although other non-fundamental regularities might be explained, e.g. by subsumption). Nomicity can then be introduced in LM through a standard Best System Account; given a certain regularity *r* (standardly expressed as a universal quantification), it is a law that *r* because "r" belongs to the deductive system with the best combination of simplicity and strength. We may then introduce a plurality of worlds, and thus distribution of qualities across those worlds as well; this allows us to treat possibility and necessity. But more importantly for our current

[14] Carroll (1994), Maudlin (2007).

[15] Lange (2009), Whittle (2009).

[16] This is the so-called "modalism", as in deRosset (2014), Forbes (1992), and Peacocke (1978, 2002); also recently discussed in Bueno and Shalkowski (2009, 2013, 2015) and Wang (2018).

[17] Explicitly: Cross (2005), Borghini (2009), and Mumford and Anjum (2011).

[18] This is the purview of the so-called "neo-Humean metaphysics", as defended by David Lewis – e.g. Lewis (1986).

purposes, a plurality of worlds, together with nomicity, gives us enough resources to define a ternary relation of comparative similarity across worlds: this allows us to account for subjunctive conditionals along familiar lines.[19] Now LM is equipped with a notion of counterfactual dependence that can account for causation. Finally, with this wealth of resources we can finally explain dispositions; in LM, this may take the form of a "refined" conditional analysis of dispositions, as in Lewis (1997), which requires counterfactual conditionals and causation to function.

PM provides a completely different rearrangement of the nomic family; amongst the distribution of qualities at the bottom of the explanatory hierarchy, there now are special dispositional ascriptions. Dispositionality is an explanatory primitive at least in the sense that some dispositional ascriptions are not *explananda*. However, some regularities now are, viz. the non-accidental ones.[20] The basic idea to be developed by PM is that, at least in some cases, reality displays such-and-such patterns of regularities because things are so-and-so disposed. Dispositions also account for laws of nature and subjunctive conditionals; finally, from a power-based account of subjunctive conditionals, we can explain certain necessities and possibilities, perhaps along the lines of Jacobs (2010).

How are we to choose between LM and PM? Disagreements about principles (e.g. Humean Supervenience) risk to turn dogmatic, and any debate conducted at a similarly generic level may be too abstract to be of any value. It goes without saying that this is just a sketch of LM and PM: there are different ways to fill in the details with proper explanations, and thus obtain mature metaphysical theories to be properly evaluated. In the following chapters, I will submit one of such ways to fill in the details. Yet certain things are already emerging which deserve our attention. First and foremost, LM works on a much sparser explanatory basis than PM. As primitive distribution of qualities and regularities require no special conceptual machinery beyond first-order logic to be expressed, there is no additional expression of the nomic family for us to apply our $\mathbb{S}$ operator to. All subsequent analysis are "reductions" at least in the loose sense that nomic and conditional conceptual machinery is explained away in a way that does not presuppose further expressions within the nomic family. For PM, on the contrary, there's dispositionality at the bottom. In fact, things might be ever worst for them: depending on the kind of explanation provided, it's possible that there will be further primitives for PM to take into account; maybe the various explanations provided by PM do not simply require things being so-and-so disposed to function; perhaps they will need dispositions as properties, in the sense that property-talk or instantiation-

[19] Standardly, this kind of account of subjunctive conditionals is presented as a semantic analysis, rather than an explanatory endeavour. I take this semantic ascent as a way to pursue a certain project all the while hiding a certain embarrassment in matters of metaphysics, perhaps a heredity of the linguistic turn in analytic philosophy. In my opinion, the question "why is 'if kangaroos had no tails, they would topple over' true?" hides a more direct question: "why is it the case that if kangaroos had no tails, they would topple over?".

[20] This is where we first encounter the peculiarly non-Humean notion of "governance". I will come back to this in Chap. 7.

talk will be used; similarly, many friends of dispositions make liberal use of essence-talk in this context (here is a widespread claim amongst dispositionalists: "...in virtue of the essence of dispositional properties..."). If such expressions are indeed to be found at the bottom of the explanatory hierarchy, and are thus taken to be joint-carving by the realist about dispositionality, then the structure of reality posited by PM will be vastly more complicated than the one posited by LM. All else being equal, LM would therefore function from a sparser explanatory basis than PM, and would therefore be more virtuous; if that were the case, such theoretical virtues of LM (vis-à-vis those of PM) would lend some degree of credibility to the idea that the dispensability of disposition-talk in such a framework is metaphysically significant (it would also, indirectly, lend credibility to other components of that explanatory framework, e.g. the primitiveness of the distribution of qualities).

Luckily for the friend of powers, we are not at that point yet: for all else is *not* equal. Two things must be noted. Firstly, LM and PM have been merely sketched, and it is far from clear that their supposed explanations will function as intended, or that they will not need further resources to avoid potential counterexamples.[21] Secondly, when that has been cleared, one should evaluate and compare the explanatory power of the two theories. Can the same (intuitively true) subjunctive conditionals be accounted for? Which laws of nature can be more easily recaptured? Although details might differ, LM and PM fare naturally better in some regards rather than others.[22] This explanatory power of the theory must be confronted, amongst other things, with the simplicity of their primitives.

At this point, various arguments which are classifiable as Arguments from Indispensability may be submitted, by the supporter of PM, in favour of the perspicuity of dispositional locutions – emphasizing some theoretical virtue of PM over its rivals.[23] It's a long way to go, of course; on the one hand, one needs mature enough theories to be compared – not to mention a somewhat unified meta-

[21] As far as I know, no step in LM, as I have presented it, is entirely uncontroversial. There are troubles with the Best System Account of laws as regularities, with the counterfactual analysis of causation, and with the conditional analysis of dispositions (more on that later). PM is equally incomplete.

[22] E.g., a conditional-first power-based approach to modality such as the one to be found in Jacobs (2010) is best suited at treating certain classes of subjunctive conditionals, such as causal counterfactuals – yet patently not all subjunctive conditionals are causal counterfactuals. As for laws, there are symmetry and conservation principles that friends of powers have famously struggled to countenance, and which are *prima facie* treated much easily in other frameworks.

[23] The reader may think that when understood as an argument from explanatory indispensability is a bit anomalous; usually, an argument from explanatory indispensability requires a fixed *explanandum* which can be explained more or less efficaciously by this or that theory. Here the situation is a bit complicated, as different arrangement of the nomic family require that one's theory *explanandum* is another *explanans*. There is however a fixed pool of phenomena to draw *explanans* and *explananda* from – viz. the so-called "nomic family" itself, so I think that this counts as a kind of explanatory indispensability in a broader sense. Dispositions can be said to be explanatory indispensable in the sense that, when accounting for the nomic family of notions in its entirety, they cannot be swapped for other *explanans* without some significant theoretical loss.

metaphysical framework allowing them to be compared. On the other hand, one must always keep in mind that no verdict can be reached by simply comparing one's favourite theory with an adversary: many comparisons must be considered (this was yet another case in which my comparison between LM and PM was an oversimplification). Therefore, developing such arguments is more likely to take the form of an ongoing process of ever-increasing corroboration.

5.2.2 Dispositional Reality as Explanatory Dispositionalism

This strategy is meant to vindicate the usual *modus operandi* of dispositionalists, many of which focus on a somewhat self-contained development of their theory, and only in a second stage operate theoretical comparisons. It also sheds some light on what kind of theoretical enterprise realism about dispositionality really is – viz., a primarily explanatory enterprise.

Allow me repeat, once again, that debates on the reality of dispositions are usually pursued under the influence of the Ontological Temptation, which turns them into debates concerning the metaphysical status of certain entities – usually the existence of dispositional properties, their essence or fundamentality, or a combination thereof. If the temptation is strong enough, arguments in favour of dispositions take the form of Lewis-stye cost-benefit analysis in favour of admitting dispositions in one's inventory of the world; and so, embracing realism about dispositionality becomes an exercise in ontological inflation – with all the pros and cons that come with it. To give a less abstract example, consider how the "actuality" or "naturality" of dispositions is usually considered an important benefit in selecting dispositions as one's preferred ontology of modality.[24] Easy epistemic access to dispositions is also considered another benefit of dispositionalist ontologies of modality, vis-à-vis the inaccessibility of Lewis-style possible worlds (and perhaps *ersatz* worlds too). In short: dispositions are inherently virtuous ontology, in the study of modality and other related fields.

Things are different in the approach we are pursuing, however. As the "joints" of reality that metaphysics is trying to explore are not reified, we should not expect any metaphysical analysis of the nomic family to uncover dispositional entities, or nomic, or modal entities. Therefore, the competing metaphysical theories introduced

[24] Dispositional treatments of modality are supposedly actualist not only because they accept the actualist maxim that everything is actual, but because they are supposedly able to function without using Leibnizian biconditionals as bridge-principle between the usual dual modalities as sentential operators, and quantification over worlds; see Eagle (2009), Contessa (2010), and Vetter (2011, 2015). The commitment to *metaphysical naturalism* – viz. roughly put the idea that only the material or the physical is real– is a bit more elusive, but equally pervasive; perhaps it emerges more naturally in contrast; see Dumsday (2015) and Giannini and Tugby (2020).

in the previous subsection should not be understood as pitting ontological postulates against each other (e.g., powers vs. possible worlds, subjunctive facts vs. robust laws of nature, etc); they should rather be understood as providing competing explanations concerning a problematic range of appearances, such as dispositionality, modality, nomicity, and the like – most importantly, "explanations" in the deflated sense explored in Chap. 3, viz. as offering answers to non-causal why-questions concerning metaphysically puzzling truths of ordinary languages such as English.

This explanatory approach to dispositionalism is in fact a close relative to the kind of approach to be found in Azzano (2019) and Vetter (2020). In particular, my approach is similar to Vetter (2020), to the extent that we both use a metaphysically deflated notion of explanation, according to which to explain a disposition – or to explain with a disposition– is to provide a completion for "...because..." sentences in which one side we find a dispositional ascription, without the need of reified *explanans* and *explanandum* in the form of a dispositional property, a dispositional property instance or fact, or the "essence" or "nature" of a property, or anything of that sort. In other words: if this theory of Dispositional Reality is a kind of power metaphysics without powers, it should rather be understood as an explanatory endeavour, in the deflated sense seen above.

On such an approach, many of the ontological debates seen before, and comparable benefits brought by one's candidate, are similarly deflated. E.g., there is no debate to be had about whether powers *qua* entities, or possible worlds for that matter, are "naturalistically acceptable entities", or something of the sort. Similarly, there are no modal or dispositional entities whose epistemic accessibility we can investigate and evaluate – in fact, it seems to me that to wonder what our epistemic access to Dispositional Reality rests on a categorical mistake (although there of course remains the question as to how we can know whether disposition-talk or possible-world-talk is perspicuous, which is what we are presently trying to answer).

A quick glance at the history of the rehabilitation of dispositions in contemporary metaphysics and philosophy of science will reveal that this rehabilitation does not hinge on a shared and clear understanding of what of entities dispositions are supposed to be. Nor there is a convincingly enough argument for the existence, or fundamentality, or what have you, of powers *as* entities. Dispositionality, so to speak, has been put forward as an explanatory hypothesis: and more precisely, it was submitted as the cornerstone of a crucial inversion of the explanatory hierarchy of the nomic family (viz.: what if we didn't try to explain dispositions through counterfactuals and other modal notions, and do the contrary?). My methodology is meant to recapture this crucial fact: the beating heart of power metaphysics lies in its applications. And it is such applications – viz. non-causal explanations of nomic and modal phenomena attempted with the help of powers—that the rest of the book will focus on.

Before moving on, a clarification. Some metaphysicians who are drawn to dispositions indeed do so not because of any alleged intrinsic virtue, either ontological or metaphysical, but by virtue of the metaphysical heavy-lifting that dispositions can do in other areas of metaphysics; in this sense, recent grounding theories of powers crucially resemble positions such as the truth-making theory of powers, as in

Jacobs (2010). But there is a crucial difference between those accounts and the present one. According to the present version of explanatory dispositionalism, to put forward realism about dispositionality consists in attempting an explanatory endeavour; but this does not mean – as some grounding theorists of powers perhaps believe— that it consists in the discovery of a grounding-slash-explanatory architecture in the world; my anti-layering position put forward in Chap. 3 makes it very clear that I do not believe that any such architecture exists. This is also the main difference between the present account and Vetter's version of explanatory dispositionalism. According to Vetter (2020, p. 17, fn. 11) the dispositionalist "ha[s] to commit to the idea that there are objective explanatory relations in the world" as, without them "the entire debate does seem misguided". This is not true on my approach, at least not when it comes to the constitutive explanations we have discussed so far. There's no out-there (constitutively) explanatory structure of reality: rather, as metaphysicians we provide bridge-principles which are explanatory in an epistemic sense, as a tentative attempt to individuate metaphysically privileged ideology.

5.3 Other Arguments

Before performing this rearrangement of the nomic family in the next chapters, there is something else that I would like to discuss. Another goal of this chapter is to discuss some of the previous argument submitted in favour of power metaphysics, and evaluate them in the light of the current approach. This is not for the purpose of an arbitrary blanket exegesis of the debate, but because such a discussion can bring our methodology into sharper focus.

5.3.1 The Conditional Analysis

When trying to characterize in a catch-all fashion the modal aspect of dispositionality, I claimed that a disposition characterizes an object not as it actually is, but as it could be, in a given circumstance: in fact, dispositionality and conditionals are perhaps the most intuitively correlated phenomena in the nomic family. This association between dispositions and conditionals can be quite straightforward at times; e.g. a glass' disposition to break if struck can be characterized by saying that the glass would break if struck; a sugarcube's disposition to dissolve in water can be characterized by saying that the sugarcube would dissolve if immersed in water. Traditionally, such conditional characterizations were put forward as a veritable conditional analysis of dispositions. An instance of such an analysis could be something along the lines of:

SCA_1) the glass is disposed to break if struck iff the glass would break if struck

But how exactly does the debate around (SCA_1) and other similar biconditionals relates to our quest for a justification for realism about dispositionality? The answer is a bit complicated. To approach an answer, we might start by noticing that, as simple as it looks, this so-called conditional analysis of dispositions was plagued by massive difficulties from the very start. Originally the first difficulty was due to the fact that the analysis had been proposed before the so-called "modal revolution" of the '60s and '70s, and therefore before the popular Lewis-Stalnaker semantics for subjunctive conditionals; and yet, treating the conditional as a material conditional notoriously gave rise to problematic consequences (e.g. all unstruck things are disposed to break if struck). This elicited different reactions; while some tried to navigate these shortcomings by amending the analysis (Carnap, 1936), other took the bull by the horns and tried to offer some account for subjunctive or "contrary-to-the-fact" conditionals (Chisholm, 1946; Goodman, 1947); neither solution seemed particularly successful, which perhaps partly motivates Quine's (1973, p. 10) dismissive attitude towards dispositional ascriptions:

> I am with Carnap in not settling for definition of dispositions by the intensional conditional. Unlike Carnap, however, I am not concerned to establish the disposition idiom as a technical idiom of scientific theory at all.

Quine conceived of dispositional terms as temporary placeholders for microstructural traits that future scientific successes will make superfluous; e.g. a sufficiently clear account of the crystalline microstructure of a glass will make any reference to its fragility, or its disposition to break if struck, entirely superfluous. But even those with a less adversarial attitude were unsettled by the troubles of accounting for subjunctive conditionals; curiously enough, in this context the association with microstructural or otherwise "respectable" features of an object made dispositions more respectable than the subjunctive conditionals deployed to account for them, as in Goodman (1954, p. 43), and Quine (1960, p. 204). Yet, all in all, it was a dire situation.

One would think that the success of the Lewis-Stalnaker semantics for subjunctive conditionals, based on a ternary relation of comparative similarity between worlds would finally vindicate the conditional analysis of dispositions. Unfortunately, that is not the case: in recent literature a massive number of counterexamples were produced against principles such as (SCA_1), collectively taken as "interferences" with the dispositions' manifesting, such that the dispositional ascription would be true, but the conditional false – or vice versa. Martin (1994) first proposed a set of counterexamples to the "simple conditional analysis" (SCA, as above), a kind of interference known as "finks" and "reverse-finks"; Lewis (1997) replied with a qualified and "reformed" conditional analysis (RCA), which was then subjected to a new battery of counterexamples in Bird (1998), later called "masks" and "mimics"; by now, the literature on the subject has witnessed countless cycles of analysis and counterexamples, and no consensus on the matter has been reached: it has not been settled which, if any, conditional analysis will get the job done, nor the reasons for these failures have been systematized to the point that it can be definitively proven that no analysis ever will – the importance of this second objective being stressed in

Bonevac et al. (2011, p. 1144–1145). Perhaps close to the mark came Eagle (2009) as he pointed out that in the standard Lewis-Stalnaker semantics subjunctive conditionals, unlike strict implications, are non-monotonic, such that their truth is sensitive to a variety of contextual and environmental factors not explicitly mentioned neither in the antecedent nor the consequent; whereas some dispositions are intrinsic, at least in the sense that the truth of the relative ascription does not depend on any such environmental factor; as such, there is no hope to perfectly line up the two, disposition and conditional, in a materially adequate biconditional. Of course, for Eagle's reasoning to go through, the conditional in the *analysans* must be a subjunctive conditional in the standard sense.[25]

The link between dispositions and conditionals is a conceptually powerful tool, and I myself have exploited it when trying to characterize the world of appearance as teeming with dispositionality; yet if the conditional analysis fails, such a link will be "only [a] clumsy and inexact linguistic gestures to dispositions" (Martin, 2007, p. 19). Some perceive an air of defeat surrounding the project, perhaps best epitomized by Mumford and Anjum (2011, p. 193), according to which we should "draw a close to the Sisyphean task of endlessly producing new candidate analyses, only to see them fail and then starting all over again. Enough ink has been spilled on this doomed project already". Not everyone, however, shares this attitude (Manley & Wasserman, 2011, p. 1191); in fact, given the lack of certainty one way or the other, there are more recent attempts to make some version of the conditional analysis work.

I am not going to break new grounds on this topic; that is to say, I am not going to argue which one, if any, of the conditional analysis will get the job done, and neither I will argue that no conditional analysis will ever get the job done. Given the size of the relative literature, this is the topic for a chapter of its own, and not one which fits comfortably in this book.[26] If anything, I would like to focus on the "analysis" part of the so-called conditional analysis: much hinges on this word – and yet, it is not exactly clear what kind of philosophical endeavour is being pursued here. The objective, and the method of achieving it, is hardly ever stated; on the contrary, discussion concerning interferences counterexamples mostly focused on the extensional adequacy of the biconditional linking dispositional ascriptions and conditionals: but what is this biconditional supposed to accomplish? And if no such biconditional can be found, what would be the consequences?

Allow me to begin with something obvious: the availability of a materially adequate biconditional between conditionals and dispositions is a necessary but

---

[25] Also see Bonevac et al. (2011, p. 1106). For a very different diagnosis of the shortcomings of the conditional analysis, Mumford and Anjum (2011, p. 190–193).

[26] For an up-to-date bibliography, the reader might want to consult some of the most recent attempts, as in Choi (2006, 2008, 2009), Manley and Wasserman (2008, 2011), and Contessa (2013a, b). Choi is perhaps the only one to defend the simple, unrefined conditional analysis presented above. Lowe (2011) offers some illuminating insight on the status of stimuli. Vetter (2015), building on Manley and Wasserman (2011), criticizes the conditional analysis of dispositions from a slightly different angle.

not sufficient condition for any "analysis" of the latter in term of the former, conceptual or metaphysical as it might be; after all, biconditionals can be walked both ways, and for a variety of different purposes. Foes of the conditional analysis who believe that powers are truth-makers for conditionals, or that they provide a semantics for conditionals, as in Jacobs (2010), must also assume the biconditional. This point ties back to something I said in Chap. 3: the mere presence of bridge-principles such as the conditional-disposition biconditional is not sufficient for any progress in metaphysics, as it is not yet established whether one side is more metaphysically perspicuous than the other; even when one direction of analysis intuitively strikes the philosopher as *prima facie* more explanatory than the other, that is still not evidence enough that any genuine analysis has been achieved: a different methodology must be employed to settle matters of perspicuity.

A lack of faith in the metaphysical relevance of the conditional analysis is occasionally echoed within the literature. E.g. both Bird (2016), and Azzano (2019) suggest that the debate on the conditional analysis is completely orthogonal to the question of the realism for dispositions. Notably, Bird (2016, p. 10) states that

> [o]ne might hold that SCA is true and that is because the existence of a property with the dispositional character on the left-hand side of SCA is the truthmaker for the conditional claim on the right-hand side. Nor does the lack of any satisfactory analysis of dispositions imply the existence of ontic dispositional properties. What debates about the analysis of dispositions focus on is typically predicates such as 'is fragile' or 'is disposed to m when s'. The lack of an analysis of such *predicates* tells us nothing about which entities there are in our ontology, i.e. which ontic properties there are. [. . .] For what we need is an argument that these expressions refer to ontic properties, not to predicatory ones, and the lack of analysis does not indicate the former rather than the latter.

It seems to me that something about this passage is right, and something else is wrong, depending again on what we mean with "analysis". It is true, as I stated above, that the existence of a materially adequate biconditional says nothing about the reality of dispositions – in Bird terms, about the existence of a property with a dispositional character, or "ontic dispositional property". Bird then goes on to state that even the lack of a conditional analysis, understood as a conceptual analysis of predicates, tells nothing about which properties stand for such predicates: so whether or not there is a successful answer to interferences counterexample, no stable foothold is gained. This part is somewhat confusing to me: Bird appears to be performing an unmotivated semantic ascent by assuming that principles such as (SCA_1) above can only be read as conceptual analysis of *predicates*, even though nothing about their formulation makes that obvious. Perhaps Bird was influenced by the fact that in the heydays of the conditional analysis, e.g. Carnap (1936), that was the preferred reading; but it is far less clear that, e.g. the exchange in Martin (1994) and Lewis (1997) was entirely concerned with this kind of conceptual analysis. It is of course true that, to the extent that it involves biconditional, the conditional analysis revolves around the logical equivalence between a dispositional ascription and a conditional (namely, sentences); but it is far less clear that the proper analysis stacking on top of the biconditional has no ontological or metaphysical import. Bird is surely right about one thing however, viz. that "what we need is an argument that these expressions refer to ontic properties, not to predicatory ones" (something that, admittedly, a conceptual analysis of predicates will not achieve); yet some other (presumably metaphysical) reading of the conditional analysis might achieve just that.

As the reader of Chap. 4 will remember, Bird's shift from "predicatory disposition" to "ontic dispositions" in my framework is recasted as a shift from the truth to the perspicuity of dispositional locutions; however, the availability of a materially adequate biconditional, by itself, is not enough to ensure that moving in either direction involves increased perspicuity of the relative sentences. As I submitted, the justification of this shift from truth to perspicuity mostly hinges on the availability and (holistic) evaluations of (collections of) explanatory biconditionals: and yet, although understood in an operationalist and deflated sense, such explanatory principles are not primarily about predicates (just as they are not about facts, or properties in a heavy-weight sense). E.g., when discussing why this sugarcube is disposed to dissolve in water, I am not primarily discussing what "disposed to dissolve in water" means, just as I am not primarily discussing the existence or fundamentality of a special property corresponding to that predicate; these explanatory endeavours do not need to be recasted in other terms, either by way of semantic ascent or "ontological descent"; they are perfectly understandable on their own. And so, the availability of a materially adequate biconditional linking dispositional ascriptions and conditionals of some kind, is bound to play some role in the discussion.[27]

Thus the two issues at stake here – conditional analysis on the one hand, and dispositional realism vs. antirealism on the other– are not entirely unrelated, although perhaps the connection between them is somewhat looser than originally thought. To see this point more clearly, remember the comparison between Lewisian Metaphysics (LM) and Power Metaphysics (PM) from before. Suppose, for the sake of the argument, that after much effort you are able to overcome all interference counterexamples and arrive at a biconditional successfully linking dispositions with subjunctive conditionals. LM may then be understood as an explanatory architecture: from a primitive distribution of qualities to natural laws, from such distribution of qualities (across worlds) and natural laws to subjunctive conditionals, from subjunctive conditionals to counterfactual dependence and thus causation, and from subjunctive conditionals and causation, finally, to dispositions. But what reason is there to believe in this explanatory architecture, viz. what reason is there to believe that those bridge-principles must be read in the direction the supporter of LM is reading them? What you, as a supporter of LM, should be focusing on to motivate the adoption of your theory is the inherent preferability of LM's body of explanations, as opposed to another. We have sketched some preliminary details above, concerning this case. On that basis, there would be some justification in

[27] Although the reasoning might differ, I thus agree with Williams (2019, p. 11) when he claims that "there is more at stake [in the debate concerning the conditional analysis] than mere predication". Also see Bird (2016, p. 349) and McKitrick (2018, p. 39–41).

thinking that the biconditional linking conditionals with dispositions in fact establishes that disposition-talk is less perspicuous than conditional-talk, and thus, that dispositionality is not real. Thus the resolution of the interferences counterexamples to some kind of conditional analysis is a necessary, albeit not sufficient condition, for an argument in favour of LM.

Conversely, suppose that you are not a foe, but a friend of dispositionality, in the sense that you are a supporter of my account of Dispositional Reality. Because, as we have seen, a biconditional link between dispositions and conditions is not a sufficient condition for the rejection of realism about dispositionality, you are not particularly concerned with the fate of the so-called conditional analysis. And yet, this is not a debate you should entirely disengage with. There is a twofold connection here. Firstly, as we have seen, the existence of a biconditional is not a sufficient, but a necessary condition for an appealing version of anti-realism; and so its achievement should at least worry the dispositional realist, to the extent that it effectively allows for an alternative on the market, convincing or unconvincing as it may be. Secondly, there may be something for you, as a realist, to gain by entering this arena. If you were to effectively prove that no biconditional link between dispositions and conditional is forthcoming – perhaps because interference counterexamples really are as deep-rooted and systematic a problem as some claims– then that would strengthen your realist position about dispositions. In fact, the non-existence of this link is evidence of the claim that disposition-talk is indispensable, at least when it comes to conditionals. Of course, there may be other bridge principles between dispositions and something else (nomic talk, possibility, necessity, etc. . .) so the evidence is not entirely conclusive: it is neither a necessary nor sufficient condition for realism about dispositionality. But it surely constitutes some evidence in that direction.

In conclusion, the debate on the conditional analysis of dispositions is so far inconclusive. It is also unclear exactly how an analysis, if available in the first place, will perform as a justification for anti-realism about dispositionality (and, conversely, how its lack will perform as a justification for realism). The way I see it, there is an intersection between the two debates, albeit perhaps a more indirect one than traditionally thought. On the one hand, a successful conditional analysis – viz. as a materially adequate bridge-principle between dispositional ascriptions and conditionals, is a necessary but not sufficient condition for an anti-realist argument; on the other hand, an hypothetical failure of the conditional analysis will lend some degree of credibility to realism about dispositionality, by lending credibility to the idea that dispositional conceptual machinery is irreplaceable, at least with respect to conditionals.

Finally, a warning. The friend of dispositions should not rejoice *too much* at the failure of the conditional analysis. This, I think, is important to stress out. After all, they are supposed to argue in favour of Dispositional Reality based on the virtues of their body of explanations concerning the nomic family – and for that, bridge-principles are needed to connect the members of the nomic family. Thus, severing one of such connections makes it difficult to provide a virtuous theory. Of course, there could be other principles linking dispositions with other members of the nomic family, which, in turn, perhaps with the help of some additional machinery, could connect with counterfactuals. But, as a rule of thumb, a dispositionalist should not rejoice at the theoretical isolation of a member of the nomic family; for any such isolated member, as corresponding to true sentences as it might be could turn into a conceptual idle wheel.

5.3.2 No Humility

An epistemological argument often submitted in favour of dispositionalism is that it avoids a kind of "humility" concerning properties, their instantiation and their intrinsic nature – something occasionally also called Ramseyan humility or Kantian humility.

The argument is usually intended to support the idea that properties are "essentially dispositional", or "nothing but dispositions", and is first expressed, as far as I know, in Shoemaker (1980, p. 214–215). The idea is the following. We know which properties things have by causally interacting with them; this may take many forms, like simply watching an apple to see that it is red, touching a surface to perceive that it is rough, or, perhaps a little more elaborately, observing the behaviour of ionized oil droplets in a special experimental environment to calculate the only unknown variable – the elementary electric charge. All in all, we interact with objects and particles in our environment through the causal dispositions that they (and we) possess – such as the disposition of a rough surface to feel a certain way under the skin; through these dispositions, we acquire knowledge about how things are.

But, if a roughly neo-Humean metaphysics is correct, and thus if properties and dispositions are allowed to co-vary independently, then this disposition-based access to properties must be severely downplayed. Propertied objects behaving in a certain way might behave in a completely different way in another possible world, the same properties being associated with different dispositions; in the most extreme cases properties may completely swap dispositions all the while remaining the properties they are; assuming, as it seems harmless to do, that such properties can also completely swap their instantiation pattern, then we could conceive of a possible world exactly like ours, but in which two properties have completely taken each other's place, while remaining the properties they are: "[t]wo different possibilities can differ just by a permutation of fundamental properties" (Lewis, 2009, p. 209). By accessing and individuating properties through dispositions, in this framework we would, at best, be accessing contingent features of these properties – their genuine nature, preserved across world, forever eluding us: on this view, properties are "quiddities", at least in the sense that "there are no restrictions on the recombination of properties in metaphysically possible worlds" (Smith, 2016). When individuating these quiddities we can, at best, individuate classes of dispositionally indistinguishable properties, with any choice within the class effectively being verification-transcendent. Hence, the humility. This generates a cascade effect of curious consequences. The first one is global; for if two properties in two worlds can completely take each other's place, then it is not far off to suppose that two world-maximal collection of properties in two worlds could completely take each other's place; so we can safely assume that our best physical theories can be realized by more than one collection of these quiddities. This straightforwardly entails that such theories underdetermine the answer to the question "in which possible world do we live in?" – in fact, nothing can answer this question.

A second consequence might be even scarier; if dispositional profile is not sufficient to identify two properties across worlds, it seems safe to suppose that it won't also be sufficient to identify two properties *within* a world. So the problem is also that, within a world, there is no assurance that two properties eliciting the same behaviour in fact are identical, or at least belong to the same type. Humility in this context also is the fact that we must relinquish our alleged ability to know which properties are instantiated through causal interaction. This is a tough one to swallow. You touch a certain surface, which elicits a certain sensation. You claim that it is rough. You touch another surface, which elicits the same behaviour; you would want to claim that the second surface is also rough – but if this framework is correct, your experiment proves very little about how the properties instantiated by the two surfaces. This kind of pattern recognition is a crucial component of both our everyday life and scientific endeavour. E.g., the aforementioned Millikan's oil-drop experiment, in order to yield the desired conclusion, needs to be repeated for various droplets, under the assumption that variations in their velocity depend on the causal contributions of the *same* properties; without that metaphysical assumption in place, the experiment goes nowhere.

A natural reaction: this is pointless metaphysics! But fear not, we are told: there is a cure. We can reject this framework and instead embrace one in which properties are "essentially dispositional", or "nothing but dispositions", or perhaps a grounding theory of powers in which properties ground dispositions and thus necessitate them. In all such cases the modal independence between properties and dispositions is eliminated: our epistemological access to properties through dispositions is restored. With a little more effort, we may also restore our ability to know which properties there are through causal interaction: all it takes is to ensure that no two properties are associated the same powers; e.g. through Bird's (2007a, p. 72) "strong" variant of dispositional essentialism, or the claim that no two properties ground the same dispositions. Similar stances are indeed common in the literature.[28] This stance is sometimes presented a reaction to quidditism itself.[29]

In my opinion, the no-humility argument is not a knock-down argument, for two reasons. Firstly, because one can simply bite the bullet: one may simply accept the consequences of humility, as peculiar as they might first strike – e.g., they might accept that it is impossible to individuate properties with a finer grain than classes of dispositionally indistinguishable properties. As Lewis (2009, p. 210) famously claimed: "[w]ho ever promised me that I was capable in principle of knowing

[28] Besides Shoemaker (1980), also see Whittle (2006) and Bird (2007a, p. 78); the position even has some grip outside of the confines of power metaphysics, as in Rayo (2013, p. x–xi).

[29] See Black (2000), Schaffer (2005), and Smith (2016).

everything?".[30] What is more, humility is occasionally accepted by friends of powers as well, more specifically those who claim that, although quidditism is false – so understood as a thesis about modal permutation—, our disposition-based access to properties is inherently partial if we also accept that, for whatever reason, such dispositions do not exhausts a property's nature.[31]

Secondly, and perhaps more importantly, the no-humility argument is arguably an argument against a weaker thesis than what friends of dispositions usually assume. Let us assume, *contra* the previous considerations and for the sake of the argument, that humility is indeed an unacceptable conclusion which straightforwardly warrants a *modus tollens* for the negation of a premise. But what would that premise be? That would be what Bird (2016) calls the "modal fixity of powers", viz. the thesis that properties and disposition are not modally independent, at least in the rough sense that, necessarily, things with the same dispositions instantiate the same properties (inside a world or across worlds). After all, following Smith (2016) we defined the relevant brand of quidditism in similarly modal terms. Let us call "Modal Fixity" the above claim of modal non-independence of properties and dispositions, and let us call "Humility" the description of the aforementioned problematic situations concerning epistemic access and individuation to properties; thus "Non-Humility" as the negation of Humility, is the rejection of all such problematic situations. Thus a single argument may run as follows:

1) If No Humility, then Modal Fixity
2) No Humility
3) Thus, Modal Fixity

This *modus ponens* is simple enough. Premise (1) states what we have said above about the modal source of Humility: to reject Humility is to accept that properties have dispositionally fixed roles. As for (2), we have accepted, for the sake of this argument, that Humility is unacceptable. Is that the argument that the friend of powers has in mind? Possibly. After all, Bird (2016) considers Modal Fixity the "special ingredient" of power metaphysics – which distinguishes properly dispositional properties (those with modally fixed roles) from the others. Unfortunately, this point is severely problematic, as shown in Azzano (2019, 2020); Modal Fixity is insufficient to characterize a form of power metaphysics, formulated as it might be in terms of grounding or essence; in terms of grounding one might notice that the presence of a modal connection is not sufficient to warrant the presence of a grounding one, as grounding is usually conceived as a more-than-modal notion; similarly, such a modal connection is not sufficient to warrant the conclusion that properties have an inherently dispositional essence, given that there is a wide gulf of

[30] This stance has also been adopted in Russell (1927, p. 390), and Maxwell (1978, p. 390); more recently is has received independent philosophical motivation in Whittle (2006), Locke (2009), Kelly (2013), and Yates (2018b).

[31] Williams (2019), chapter 5.1. Williams (2011a, p. 86) explicitly embraces Lewis' claim of humility.

positions between the rejection of this kind of modal quidditism and dispositional essentialism proper, as in Smith (2016) and Azzano (2020). My favourite example is what I have occasionally called "Necessitated Categoricalism", viz. an Armstrongian theory of properties and laws in which laws of nature, as second-order relational instances, are themselves necessary as opposed to contingent – maybe exactly to deal with unpleasant consequences such as the above. But it is not the only way: if this kind of brute necessities strike you as unpleasant, you may follow Lange's (2009) adoption of primitive subjunctive facts as lawmakers, in which, allegedly, there is no primitive necessity at all. Most importantly for our current purposes, this makes the non-humility argument inconclusive.

In conclusion, Modal Fixity is not the exclusive purview of the friend of powers. Consider dispositional essentialism, viz. the claim that properties have a dispositional essence, in the sense that those dispositions that necessarily accompany it (Modal Fixity) are *also* essential to it (similar arguments could be formulated, I believe for other variants of power metaphysics). Now consider the following argument.

4) If No Humility, then Dispositional Essentialism
5) No Humility
6) Thus, Dispositional Essentialism

The problem of this argument is that (4) is obviously wrong, in the sense that, as we said before, the failure of Humility is not per se sufficient to warrant the adoption of Dispositional Essentialism. Many more options are on the table. Of course, the converse of (4) is true, which may reassure the supporter of Dispositional Essentialism. Yet the following argument is obviously invalid.

7) If Dispositional Essentialism, then No Humility
8) No Humility
9) Thus, Dispositional Essentialism

So, the argument (1)–(3) is valid and with true premises (assuming Humility really is unacceptable), but it is in support of a weaker thesis than realism about dispositions. Arguments (4)–(6) and (7)–(9) are explicitly in support of (a specific version of) realism about dispositions, but (4)–(6) has a false premise, while (7)–(9) is invalid; and this line of reasoning can be generalized for other versions of realism about dispositions, e.g. grounding-based power metaphysics. So, the no-humility argument is inconclusive. Of course, the truth of principles such as the converse of (4) indirectly corroborates the position in the sense that it allows it to avoid an unpleasant situation (assuming humility to be unpleasant in the first place – see above); but there is no argument straightforwardly linking No Humility with dispositionalism, in any shape or form.

Finally, there's one more reason to doubt the metaphysical significance of the no-humility argument; viz., that it is unclear how much the charge would stick given the current account of Dispositional Reality. Our metaphysical background does not envision powers as properties, and is in fact compatible with at least certain forms of nominalism; it is highly unclear whether on this account there would be any residual worry concerning epistemic access to properties, and their metaphysical individuation (in Chap. 4, Sect. 4.5.2.3, I wondered whether there could be any deflated sense of metaphysical individuation, without any entity to be individuated). Worries about our access to "intrinsic nature" of properties are equally unconvincing: I do not believe in properties-as-entities, and so I do not believe in any objectual essence that they might possess, nor I worry about our access to it.

Some issues, however, remain. For one, consider this. A peculiar feature of the non-ontic notion of structure employed so far is that our joint-carving operator $\mathbb{S}$ can be applied to expressions of any syntactic category – which means that we should not expect difference in syntactic category to be underlined by deep metaphysical differences; e.g. assuming that the relevant expressions are joint-carving, perhaps the same aspect of reality is revealed by the expressions "particle A has charge *e*" and "the charge of particle A is *e*". Thus, although I revealed by nominalist leanings in previous chapters, I am perfectly comfortable quantifying over spins, charges, and so forth. Remember that $\mathbb{S}$ was meant to take the place of the all-too-ontic and quasi-referential notion of correspondence, which unjustly underlined the metaphysical importance of certain expressions (e.g. referential terms) over others (e.g. predicates); but if both expressions reveal aspects of reality, then there's little reason to think that we can quantify over one but not the other. In fact, assuming that both above sentences on particle A and its charge are true, it seems to me that "there's something that particle A has" and "there's something that is *e*" quantify over the exact same "thing" – the charge of the particle. Yet one quantifies at the first-order, and the other at the second-order.

This is a long-winded way of saying that my account of Dispositional Reality allows for first-order quantification over properties that sanctiones certain forms of humility; it is perhaps the case of Lewis' (2009) properly "Ramseyan" form of humility. Consider a profoundly idealized scenario in which we are able to identify a special sentence $\phi(t_1 \ldots t_n)$ of which all sentences of the "true and complete final theory" (Lewis, 2009, p. 205) are logical consequences (lucky us!). Now assuming that "total swaps" such as the ones above are possible (viz. two properties at two world completely exchanging dispositional and instantiation profiles), then it will be the case that $\exists x_1 \ldots \exists x_n \phi(x_1 \ldots x_n)$ is multiply realized – in the sense that more than one collection of properties, in more than one world, satisfies $\phi(x_1 \ldots x_n)$. This generates the uncomfortable consequence that there's little evidence, theoretical or observational, that one of such collections, as opposed to another, actually satisfies $\phi(x_1 \ldots x_n)$. After all, "[b]ecause [$\phi(t_1 \ldots t_n)$] is complete, it is capable of accommodating any evidence we could possibly gather, but [$\exists x_1 \ldots \exists x_n \phi(x_1 \ldots x_n)$] accommodates that evidence just as well, and [$\exists x_1 \ldots \exists x_n \phi(x_1 \ldots x_n)$] is true whichever realization of [$\phi(t_1 \ldots t_n)$] is actual" (Yates, 2018b, p. 463). Hence, the humility.

This kind of worry *is* expressible in my account – and similarly, the kind of modal fixity required to assuage this worry is also expressible. This only requires, with some simplification, that each special predicate of this "final and complete theory" $P^T(x)$ is necessarily and exclusively paired with a certain dispositional profile d(x); in other words:

$$MF)\Box\forall x\left(P^T(x) \leftrightarrow d(x)\right)$$

I assume (MF) to be schematic for place-holders "P^T" and "d"; the box is supposed to express metaphysical necessity. Left-to-right: necessarily, everything that is P^T also has a certain dispositional profile; right-to-left: necessarily, everything that has certain dispositions, it also has a certain property P^T. To the extent in which the various dispositional roles d(x) ultimately contribute to codify the content of ϕ, MF entails that $[\exists x_1 \ldots \exists x_n \phi(x_1 \ldots x_n)]$ is not multiply realizable.

So, should we accept (MF)? I am unsure. Most precisely, I'm not sure whether Ramseyan Humility is so unacceptable that it warrants the adoption of any principle that allows to reject it. This is something that I am not going to settle. Furthermore, again, one must keep in mind that even if Ramseyan Humility was in fact so unacceptable, the acceptance of (MF) does not *per se* constitute the acceptance of any brand of power-based metaphysics or Dispositional Reality: (MF) is not sufficient to establish any priority (grounding/explanatory/etc etc) of the dispositional side on the other, or vice versa.

In conclusion, while the acceptance of (MF) might be a nice bonus for the supporter of Dispositional Reality, under certain assumptions concerning the unacceptability of Ramseyan Humility, does not straightforwardly offer an argument in favour of it.[32]

5.3.3 *The Argument from Science*

Finally, we approach the so-called Argument from Science (for brevity, AS). If there was ever a direct argument in favour of dispositions, that would be it. Its importance in the debate can hardly be overstated; according to Mumford (2006, p. 471) "the whole debate between Humean and non-Humean metaphysics might rest on [its] viability".[33] I believe this to be an exaggeration, if anything because dispositionalism – the position advocated for by AS–, hardly exhausts the spectrum of non-Humean positions (this was perhaps the moral to be extracted from the previous section: for our purpose, we can understand a non-Humean position as any position which entails Modal Fixity). But never mind that: what interests us today is whether we can find a genuine argument in favour of dispositionalism in AS. My answer, which depends on the adoption of my framework for Dispositional Reality, is no.

[32] In the next chapter I will deal with the relation between scientific predicates and overtly dispositional predicates, and I will shed some light on whether we can take a principle such as (MF) as being just a strict entailment or something more (e.g., an explanation). If not, the supporter of Dispositional Reality who endorses (MF) still owes us an answer: whence the necessary connection between P^T and d?

[33] For an extensive and up-to-date bibliography on the issue, see Williams (2011a).

5.3.3.1 Presentation of the Argument from Science

AS is best understood as a reaction to a certain train of thoughts from the foe of dispositions. Following the tradition, let us call the "causal basis" of a disposition the property that is causally responsible for its manifestation (if one prefers to think about causes and effects as events, one should be able to easily rephrase the notion accordingly). According to the foe of dispositions, ordinary dispositions such as a glass' disposition to break if struck, cannot be the causes of their manifestations, at the pain of producing spectacularly uninformative causal explanations such as: the glass broke because it was disposed to break if struck. This is the famous *virtus dormitiva* objection,[34] which suggests the idea that dispositions are equipped with distinct causal basis; in many cases, scientific investigation provides such basis: e.g. a glass' disposition to break if struck has its causal basis in the crystalline microstructure of the glass itself, a complex collection of properties and relations of the glass' most minuscule components; and the same goes for most ordinary dispositions. Thus, the foe of dispositions may argue that all dispositions are "grounded" in the sense that they possess distinct causal bases.[35]

It's unclear what kind of anti-realism towards dispositions, amongst the ones presented so far, this proposition might entail – perhaps the thought is that dispositions, made causally impotent by their causal basis, would be made an expendable epiphenomenon (Mumford, 2006, p. 478); more generally, the idea is that microstructural causal bases of dispositions, rather than dispositions themselves, provide the causal or scientific explanation of that disposition's manifesting; on this conception, dispositional terms are but temporary place-holder terms for those mechanisms and portions of that causal history that still we do not fully understand, post-it notes for future discoveries (Quine, 1973, p. 10).

This train of thought is opposed by the claim that some dispositions are fundamental, e.g. McKitrick (2003, p. 350). With some approximation, in this context a "fundamental" property simply is the property of a physical entity without proper parts, presumably particles studied by fundamental physics (Williams, 2011a, p. 72).

[34] For the record, I believe that the correct resolution of the *virtus dormitiva* objection lies in Mumford's (1998, p. 139–141) observation that causal explanation is finer-grained than causation; also see Ruben (1990, p. 160). Settling this point is not crucial for the fate of AS.

[35] See Prior, Pargetter, and Jackson (1982). It is not always clear, within the debate on AS, in what sense a property "grounds" a disposition. In the standard use of the expression, as in McKitrick (2003), a property "grounds" a disposition by being the causal *explanans* of its manifestation – thus its causal basis; but in some cases, e.g. Ellis and Lierse (1994, p. 31–32) grounding properties does not just provide a causal explanation of a disposition's manifestation, but an explanation (whether causal or not, is unclear) for the continuing existence of the disposition itself. Similarly, Bauer (2011, p. 81) states that "[t]he grounding property serves as the causal basis for fragility manifesting [. . .], and also grounds the continuous instantiation of that token of the property of fragility when it is not manifesting". This curious oscillation in the meaning of "grounding" within the context of AS has not, in my opinion, been sufficiently investigated, as is the relation between this notion and the more standard notion of grounding. I leave this matter for another time, as the problems I have with AS do not hinge on this ambiguity.

Personally, I do not believe that a necessary condition to be a realist about dispositions is to believe that some of the properties of fundamental physics are dispositions – that would be to indulge too much in the "imperialist tendencies" proper to a certain kind of microphysical reductionism in philosophy of science, to borrow a colourful expression from Cartwright (1999). But never mind that. Whatever its significance, friends of dispositions have challenged this train of thought, on two grounds. On the one hand, we can notice that the foe of dispositions requires not only that all dispositions are provided with distinct causal bases, but with *non-dispositional* causal bases – something which is difficult to dictate *a priori*, and is very much dubious (Ellis & Lierse, 1994, p. 31). On the other hand, the success of this strategy appears to be undermined by the current state of physics.

This is where we encounter AS. With some approximation, AS involves a Scientific Premise (SP) and a Philosophical Premise (PP), which together lead to a Conclusion (C); summarily presented, AS takes the following form:

SP) There are propertied entities with no underlying microstructure
PP) At least some of the properties possessed by such entities are dispositions
C) Thus, some dispositions have no causal basis (viz. "ungrounded" dispositions)

The conclusion C is meant to illustrate that the aforementioned reasoning of the foe of dispositions breaks down in fundamental contexts: dispositional properties possessed by structureless particles cannot receive the kind of microstructural reduction available for ordinary dispositions such as the ones considered so far; in these fundamental contexts, there's no underlying structure that can be causally responsible for the manifestations. So, we are left with dispositions.

Is AS a sound argument: that is to say, are the premises credible? Do they lead to the conclusion? There are reasons to give a negative answer to both questions; but my reason for doubting AS is that I believe that PP, the Philosophical Premise is wrong – or, more precisely, we have not yet collected enough evidence to believe that it is true. But more on that later: let us analyse the components of AS.

SP simply consists in the *a posteriori* experimental observation that some fundamental contexts display no underlying microstructure; some fundamental particles like electrons or quark are "point-like", not necessarily because they take up no space (in fact, one shouldn't think about such particles as tiny objects at all), but because they reveal no internal structure, at least at the energy levels that we are currently able to probe. This makes SP a tentative at-the-best-of-our-knowledge type of claim, and as many others like it, it is naturally open to future refutations. But as experimental knowledge begins to pile up, there are reasons to think that such a refutation is not forthcoming. According to Molnar (2003, p. 133):

> Ever since Rutherford had found 'bombardment' as a method of probing for subatomic structure, every significant increase in the level of projectile energy has led to discoveries of new, deeper levels of structure. But now there exists a strong body of evidence, from experiment to theory, to show that this history will not keep repeating itself. Collisions have been produced in accelerators with energies over a hundred thousand times above the level at which new structures could have been expected to reveal themselves if history was going to repeat. Yet electrons and quarks continue to come out as point-like, structureless entities.

As far as premises go, PP is *prima facie* a credible one: in fact, the *a posteriori* status of SP gives us reasons to think that these ungrounded dispositions aren't merely possible, but are *actually* instantiated in fundamental contexts – which makes AS that much more dangerous for the foe of dispositions.

Now we move to the philosophical premise, PP, according to which many properties instantiated by such structureless entities, such as mass, spin, or charge, are dispositions; in fact, Blackburn (1990, p. 63) makes the stronger claim that "science finds only dispositional properties, all the way down". But there's no need to be that radical. Suffices to notice that many properties of structureless entities, such as mass, charge, spin and the like are characterized in dispositional terms, viz. as dispositions, or propensities, or capacities to react in certain circumstances in a given situation. Inertial mass is the supposedly intrinsic capacity to resist acceleration by a given force, whereas gravitational mass is understood as the capacity for an object to generate a gravitational field; electric charge can similarly be understood as the capacity to generate an electric field. A particle's spin, represented mathematically as a vector, is but a capacity to contribute to the particle's total angular momentum. Indeed, the very notion of "matter" as opposed to "anti-matter" can be understood, as a difference in sign, as the possession of the opposite charges that intrinsically dispose particles and antiparticles to annihilate each other. Such features do not provide any insight on a supposedly internal make-up or structure of the property, nor do they provide any other qualitative insight – in whatever sense of the word "qualitative"; they merely describe the particles as they are disposed to act in certain circumstances. According to PP they are, in other words, dispositional properties; in combination with SP, we can conclude that they are ungrounded dispositional properties.

Does C follow from SP and PP? There are good reasons to believe that it does not. More precisely, it may not be the case that dispositions of structureless particles are necessarily ungrounded: the grounding base might not be constituted by an internal microstructure of the particle, but by something else, be they "fundamental symmetries" as in Psillos (2006, p. 152–154), or "properties of the universe itself" as in Harré (1986, p. 295) – the latter option, occasionally called "ultra-grounding", being recently vindicated by the development of one of its instance: the extrinsic grounding of mass by the Higgs field (Bauer, 2011). The jury is still out as to whether properties of point-like particles are ungrounded, and most importantly, this might not be an issue for philosophers to settle (Psillos, 2006, p. 154).

This is a fair observation. But my concern with AS is a properly philosophical one: it lies not with the logical implication of C from the premises, nor with SP, but with the philosophical premise itself, PP, according to which physics tells us that the properties of fundamental particles are dispositions.

As hinted before, it is not always clear what result AS is trying to achieve, viz. what kind of anti-realism about dispositions AS is trying to defuse; but whichever it is, it would appear to be well within the grips of the Ontological Temptation: in other words, AS engages with realism and anti-realism concerning dispositions as an issue concerning the metaphysical status of certain entities: maybe the fundamentality, or ungroundedness, or lack of causal bases of some dispositional properties. But this is not how our investigation concerning Dispositional Reality has proceeded so far. What AS *does* establish, without the shadow of a doubt, is that dispositional locutions can appear in the descriptions of the smallest and most fundamental interstices of nature. But what reasons are there to believe that such descriptions are not merely true, but metaphysically perspicuous?

Allow me to clarify: I am not trying to dispute the very general claim that the direction of scientific investigation also constitutes a direction of increased perspicuity. Even without an explicit defence of this claim, it would be weird indeed if I argued for the existence of a philosophical methodology that leads to increases in perspicuity (as I did) without also believing that a scientific methodology also existed – whether the same methodology or not, I will not say– leading to similar results; for that would straightforwardly lead to the self-defeating conclusion that if our interest lies in the structure of reality, we should pursue philosophy *as opposed* to science. This is not what's happening here.

My point is another: that the direction of scientific investigation is not sufficient to establish the perspicuity of dispositional talk, specifically. The reason is very simple: dispositional locutions can be used to characterize the properties of fundamental particles; but they don't need to be. Other characterizations are available which are just as the valid for the purposes of scientific investigation, employing other members of the nomic or modal family. This is not just true of fundamental physics, incidentally: the point easily generalizes to other scientific contexts. E.g., a dispositionalist may feel emboldened by the fact that in the Brønsted-Lowry theory of acids, an acid can be defined as a substance disposed to donate protons in an acid-base reaction, whereas a base can be defined as a substance disposed to accept protons in such a reaction. Yet we may offer mutual definitions through subjunctive conditionals as well: we may define an acid as a substance that, if it were to react with a base, would donate a proton; and a base as a substance that, if it were to react with an acid, would accept a proton. In fact, both these formulations involve some degree of grammatical torture, the one usually performed by philosophers who wish to put their preferred expressions in the scientist's mouth; usual formulations of the Brønsted-Lowry theory deploy 'can'-clauses, or talk of habituals: an acid is a substance that can donate a proton, or that donates a proton in an acid-base reduction reaction.

There may of course be bridge-principles connecting these formulations, viz. bridge-principles connecting dispositional talk with subjunctive talk, can-talk, talk of laws of nature and the like; these are familiar items for the philosopher. But, once again, when it comes to such bridge-principles there's no hope of reading the direction of explanation in a vacuum, thus discovering which formulation is more metaphysically perspicuous is a difficult task. More importantly, it is not a task to be performed by science: from a scientific point of view, the difference between the above formulations regarding the Brønsted-Lowry theory is nil; a scientist may accept the dispositional formulation simply because, for their point of view, very little depends on the choice between characterizations employing various dwellers in the nomic/modal family. Talk of dispositions, or powers, or capacities, can be freely used in our everyday life and scientific talk: we can say that gold is disposed to (easily) deform under compressive stress, or that hydrogen peroxide has the power to oxidize other substances, or that liquid neon has an extraordinary cooling capacity; we have also seen that dispositional talk can be employed in a metaphysically innocent way, by focusing on its truth rather than its perspicuity. That this kind of dispositional talk can be easily swapped for counterfactual or nomic talk is a red flag that we are dealing with metaphysically innocent dispositional talk, where the joint-carvingness of dispositional expressions is not on the table. Science is silent in this respect, e.g. it is not a consequence of the Brønsted-Lowry theory of acids and bases that dispositional ascriptions are perspicuous – not because the theory does not belong to "fundamental physics"— but because such dispositional locutions do not indispensably appear in it. When it comes to the metaphysical modelling of the nomic family, the Brønsted-Lowry theory of acids and bases tells us nothing.

The same goes for the dispositional characterizations of properties in fundamental physics, such as mass, charge, spin, and the like. A peculiarity of the properties of fundamental particles is that the characterizations provided through items of the nomic/modal family, which characterize particles as they can behave or would behave under certain circumstances, or as they are disposed to behave, or as they behave habitually – in short, in terms of their behaviour in a bigger system– is the only kind of characterization available. This may very well be "an inescapable consequence of the methodology" (Williams, 2011a, p. 78) of the scientific practice in microphysics, whose subject of enquiry is a quasi-inaccessible region in which properties can only be indirectly detected by observing their contributions to measurable effects (think about Millikan's oil drop experiment again). For this reason, it is dubious whether such nomic/modal characterizations require to be taken seriously from a metaphysical point of view, and even more dubious than one of them in particular, at the expense of others, should.[36,37]

As in the case of acidity, scientific theories in microphysics would appear to be insensitive to the choice in metaphysical modelling that would favour one formulation to another. This is a job specifically for the metaphysician, to be achieved along

[36] Williams (2011a, p. 78) claims that "[a]s this dispositional characterization is compatible with ontologies that countenance baseless dispositions and those that require that all dispositions have categorical bases, microphysics is ontologically innocent". Williams' notion of "ontologically innocent" dispositional talk is similar to the notion of "metaphysically innocent" notion of dispositional talk available to foes of powers I developed in Chap. 4.

[37] That this kind of behaviour-based characterization of fundamental features of particles is an "inescapable consequence of the methodology" also has another consequence: if there is a non-dispositional grounding base, it will forever elude our grasp. But even if that were true (the possibility of the so-called ultra-grounding may disprove it – see above), that wouldn't be a knockdown argument. This is where the no-humility argument intersects AS; but, as before, this is not a very compelling consideration, given that humility is occasionally accepted by dispositionalists themselves, as in Williams (2011a, p. 86).

the lines sketched in previous parts of this book.[38] And according to these guidelines for a methodology in metaphysics, indispensability is a (defeasible) sign of perspicuity. Unfortunately, dispositional talk can be used to describe properties of fundamental particles, but it is highly dispensable. It seems to me that the burden of proof is on anyone who wants to claim that properties such as mass, or charge, or spin, *require* to be characterized dispositionally for the theories they appear in to have the theoretical virtues they have: a tall order indeed, if not for the simple fact that such theories are not usually formulated in the natural languages in which dispositional locutions usually appear. At any rate, this is far more than what can be proven by AS on its own.

To clarify: that dispositional talk is entirely dispensable in scientific discourse is not a sign that there is no Dispositional Reality, but that the supporter of AS is wrong in thinking that science can this matter. There are many pieces of ideology that may turn out to be explanatory indispensable in our best scientific theories, and thus maybe parts of the structure of reality: but dispositional expressions are not amongst them.

5.3.3.2 Inter-level vs. Intra-level Explanations

It could be useful at this point to draw a distinction between two kinds of explanation that dispositions might receive.[39] We are still dealing with the kind of constitutive explanation introduced in Chap. 3, in the operationalist and thus metaphysically deflated understanding of the notion. Thus, we might obtain an explanation of a disposition in the sense that we might want to fill in the blanks of the following sentence:

10) The glass is disposed to break if struck because...

There's (at least) two different kinds of explanations that might complete (10). The first one, the one widely discussed in the AS, is a usually microstructural explanation, and might mention properties and relations of parts of the glass; some of those properties might in turn involve other dispositions. This is what we will call an inter-level explanation. E.g.:

[38] See Williams (2011a, p. 79, fn. 15). To make another example, Mumford (2004, p. 4–8) argues that the choice between descriptive and governing laws is a piece of metaphysical modelling of the nomic family whose resolution does not depend on any scientific result: this is a metaphysician's job. I agree with Mumford's assessment, and I think that the same goes for the reality of dispositions: it cannot be settled by AS, viz. the Argument *from Science*: this too, is a metaphysician's job.

[39] This distinction was suggested to me by Barbara Vetter (personal communication).

10a) . . .because the glass has a crystalline microstructure

The second kind of explanation involves other features of the disposed object itself, in this case the glass. This is the kind of explanatory bridge-principle encountered in the discussion of the conditional analysis. So, e.g.:

10b) . . .because the glass would break if struck

This is what we will call intra-level explanation. (10b) is not the only intra-level completion of (10) – perhaps another one will mention a categorical property and a law of nature, or some other non-microstructural analysis of dispositions.

Inter-level and intra-level explanations are not meant to be two distinct kinds of explanations, like causal and mathematical explanation; they are rather two variants of the notion of explanation considered so far, an extension of the constitutive explanation which we encounter in our tentative efforts to move from less perspicuous talk to more perspicuous talk. We may want to visualize the situation as follows: the foe of powers believes that dispositional ascriptions such as the one in (10) are merely true; in order to prove that, they have to provide, and evaluate, their own metaphysical theory in which such ascriptions are explained away. To explain away a dispositional ascription, they can perform a "vertical move" – an inter-level explanation–, or an "horizontal move" – an intra-level explanation. It is crucial to understand that from the point of view of their availability, there is a crucial difference between a vertical move and a horizontal move. A vertical move, an inter-level explanation, is provided by either empiric or scientific investigation (e.g., one does not discover that fragility is due to crystalline microstructure from the metaphysician's armchair); such explanations depend on concrete scientific advancements, and their availability vary on a case-to-case basis. But horizontal moves, intra-level explanations, are immediately available, provided bridge-principle schemas such as one formulated in terms of the conditional analysis: to provide such schemas is the purview of the metaphysician – albeit, as we have seen, their value as a guide to perspicuity is tricky to individuate.

This leave us with a clear understanding that AS is not a make-or-break argument for or against realism concerning dispositions: in other words, one does not establish the presence of Dispositional Reality by checking whether there are "powers all the way down" or "bare powers". What AS *does* successfully establish, if anything at all, is that the foe of powers cannot perform vertical moves indefinitely: if the current state of physics is of any indication, at some point inter-level explanations are no longer available. But as we have seen in the previous section, this is not enough to establish that there is Dispositional Reality, because *horizontal moves are still available*, at every level. Furthermore, as horizontal moves are the purview of the metaphysician, it is the job of the metaphysician, who supports Dispositional Reality, to argue that such an horizontal move, albeit possible, is not theoretically desirable. Viz., it is the goal of the metaphysician to show that bridge-principles linking dispositions with other members of the nomic family ought to be read with one explanatory direction, as opposed to another. In other words, ensuring an argument in favour of Dispositional Reality along the lines sketched in this chapter, requires disqualifying both vertical and horizontal moves when explaining away dispositions – either as outright not available, or as theoretically expensive moves.[40]

5.4 Conclusions

The main takeaway from the discussion concerning the Argument from Science is that to discover whether there is Dispositional Reality, viz. whether dispositional talk is metaphysically perspicuous is a philosopher's job, rather than a scientist's.

Talk of dispositions, while commonplace in science, does not appear to be indispensable and may be substituted with something reflecting a different metaphysical model (e.g., robust laws, primitive subjunctive facts, etc...). By itself, the availability of these different models, and bridge-principles linking them, is not metaphysically significant. For example, we have seen that the mere availability of biconditional linking dispositions with conditionals – even if it were to successfully solve all interferences counterexamples—is not sufficient to dictate direction of explanation, and thus does not give us any reason to think that disposition-talk is more perspicuous than conditional-talk. Something *more* needs to be said. According to the loosely neo-Quinean methodology proposed in this chapter, to score a point in favour of Dispositional Reality one must argue that dispositional talk is *preferable*, or explanatorily indispensable in our best possible accounting of the nomic family. This requires a holistic evaluation of metaphysical theories as bodies of explanations: later in this book we will see some of these explanations.

[40]Relatedly, one might wonder whether the viability of AS is a necessary, albeit not sufficient condition for realism concerning dispositions – as it consists in a way to prevent the foe of dispositions from operating a vertical move. I think that the ineliminability of dispositional talk by way of inter-level explanations is indeed a necessary condition for belief in Dispositional Reality, but given what we have said so far, it is not a very substantial requirement. Any scientific theory is going to describe its properties through a behaviour-oriented characterization that is particularly suitable to be treated by elements of the nomic family, dispositions or otherwise; all scientific theories so far have done it, and there's little indication that the pattern will not continue. Perhaps, in William's words again, this is a metaphysically innocent and inescapable consequence of the methodology.

Chapter 6
The Extent of Dispositional Reality

6.1 Introduction

Now that we have a tentative grasp on Dispositional Reality, and the kind of explanatory theory in which claims of Dispositional Reality are embedded, we can wonder about its extent. How much of reality is dispositional, assuming that some of it is? In this chapter, we will tackle this question by addressing two interconnected issues:

(a) within the multifaceted aspects of dispositional appearances introduced in the beginning of Chap. 4, which of them ought to be selected as candidates for perspicuity? Viz., what kind of dispositional predicates ought to be taken as joint-carving by the friend of dispositions?
(b) which true ascriptions of non-overtly dispositional predicates are ultimately underpinned by Dispositional Reality? This is the distillate, within the context of my account, of the age-old question: assuming that some properties are dispositional, are all of them? Can all properties be accounted for dispositionally?

Before starting, I would like to make an important qualification. Within the context of my account, there are no dispositions-as-entities, nor there are heavy-weight properties. Whenever I will use the term "dispositions" I will simply mean true dispositional ascriptions, viz. the kind of metaphysically innocent dispositional talk detailed in Chap. 4; and similarly, when I will discuss how to "account for properties" through "dispositions", I will mean how to explain (given my operationalist notion of explanation – more on that below) non-overtly dispositional ascriptions (e.g., "is spherical") through dispositional ascriptions (e.g., "is disposed to roll").

With that said, we can begin. In Sect. 6.2 I will discuss issue (a), whereas in Sect. 6.3 I will discuss issue (b); as it turns out, I am not a Dispositional Monist, in the sense that I do not believe that Dispositional Reality is the only reality revealed by

L. Azzano, *Dispositional Reality*, Synthese Library 482,
https://doi.org/10.1007/978-3-031-52625-1_6

predication: not all properties are dispositional; but which relation is there between dispositions and other properties, remains to be seen: relatedly, the discussion of the relation between overtly dispositional and non-dispositional predication will be an opportunity to refine the explanatory notions at our disposal, thus finally converging on the much-anticipated notion of *constitutive explanation* (and conversely, non-constitutive explanation). In Sect. 6.4 I will sketch the bigger picture emerging from my discussion of (a) and (b), and finally, I will offer some concluding remarks.

6.2 Dispositional Reality and the Varieties of Dispositionality

To believe in Dispositional Reality is to believe that dispositional talk is not merely true, but metaphysically perspicuous. But which metaphysical talk? The world of appearance is teeming with dispositionality of the most disparate kinds; which is to say, there are very different dispositional ascriptions: there are dispositions, liabilities, and capacities; there are gradable dispositions; there are dispositions with or without stimulus/manifestation specification[1]; there's dispositions with only a manifestation specification; and many more. So: which of these expressions, if any, reveal Dispositional Reality?

6.2.1 Unruly Dispositions

It seems to me that at least one kind of dispositional ascriptions is not going to be metaphysically perspicuous, no matter one's take on Dispositional Reality. These are the ordinary covert dispositional ascriptions, viz. without stimulus and manifestation specification, that we encounter in natural languages such as English: gold is malleable, this rubber band is elastic, this substance is toxic, so on and so forth. The reason is that the corresponding predicates are famously unruly, in the sense that they have context-sensitive and vague, if not outright indeterminate applicability conditions, as noted in Bird (2007a, p. 20), and Vetter (2015, p. 20); e.g. in an engineering context, hardened concrete is usually considered fragile, whereas in the context of my everyday life, I don't take the plastic ruler on my desk to be fragile – at least in the sense that, unlike a glass, I'm not particularly worried about dropping it. Ordinary covert dispositions like fragility and malleability come in degrees, and which degree warrants the ascription of the ordinary covert predicate varies from context to context.

[1] Bird (2007a) calls them "overt" as opposed to "covert" dispositions.

This is also, I believe, a lesson to be learned from those colourful debates concerning interferences to the conditional analysis of dispositions. Here is a favourite example of mine: imagine a snake which injects its venom with one fang, and its antidote with another. Is the snake venomous? Is the venom deadly? In my (anecdotic) experience, ordinary speakers don't find questions such as these particularly compelling, at least not in the abstract context of a philosophical thought-experiment: were such a snake to exist, we wouldn't be particularly worried about it during our hikes, so one can easily imagine contexts in which it would not count as venomous. Chemists and biologists, however, may disagree, and in other, perhaps classificatory contexts, they may decide to rank it amongst the venomous snakes. Something similar goes for the deadliness of the venom itself. That these ordinary predicates have erratic applicability conditions is a semantic *datum* that is not for us, as philosophers, to rectify. This is yet another corollary of the metasemantic considerations put forward in Chap. 3.

This is however sufficient to determine that, although such ascriptions may very well be true, they are not metaphysically perspicuous. Given the context- and interest-sensitive nature of these ascriptions, it doesn't seem that all things satisfying such predicates are relevantly similar, nor that there are relevant differences between them and those that do not satisfy them (in fact, one and the same item may satisfy this predicate in one context but not another). To be sure, this does not mean that there is no Dispositional Reality, but that other dispositional predicates should be taken as candidates for perspicuity, and that them (plus some contextual factor) should explain ascriptions of such a badly-behaving ideology. Think of it this way: the predicate "is tall" is just as context-sensitive, and admitting of degrees, of predicates like "is fragile"; but nobody would take this as an indication that height is not real. Better regimented height-predicates are in charge of better expressing the height-portion of metaphysical structure, which in turn can account (with the help of contextual factors) for the ascriptions of the "is tall" predicate. The same goes for dispositions. In taking this stance, I thus agree with Vetter's (2015, p. 20) claim that "[b]oth vagueness and context-sensitivity are features of language, not the world". In our quest for perspicuity, we should seek for less unruly dispositional talk.

If this is the fate reserved for ordinary covert dispositional predicates such as "is fragile", there is not much hope for their overt counterparts, such as "is disposed to break if struck". After all, if the two predicates are semantic equivalents, and the former's applicability conditions are vague and context-sensitive, then the latter one's also are (we can use our usual examples: I would say, within the context of my office, that a plastic glass is not disposed to break if struck, whereas in an engineering context, we would say that hardened concrete is disposed to break if struck).

Please note that this is not to say that (maximally) determinable covert ascriptions, with a numerical variable in the stimulus and/or manifestation specification, are equally non-joint-carving. They are not. This is an interesting phenomenon to consider: the disposition to break if struck with a force of magnitude n is less vague and context-sensitive than the simple disposition to break if struck. After all, the former is possessed by anything that would break when a certain force is applied, irrespective of context. To use a non-dispositional example, “is *n* cm tall” is not as context-sensitive as “is tall”: everything, perhaps even dimensionless entities, satisfies “is *n* cm tall”, regardless of context, whereas it is only by way of context that one counts as tall. In general, maximally determinable dispositional predicates have bigger extensions than unspecified overt ones, and also more objectively defined extensions. The same is true for partially determinate ascriptions; suppose that object A is disposed to break if struck with a force of (at least) 600 N; this ascription is arguably not as context-sensitive as the ones above. Object A finds itself in Context α, which warrants ascriptions of fragility to objects disposed to break if struck with a force of (at least) 500 N. Thus, the following explanation may be put forward:

1) Object A is fragile (in Context α) because Object A is disposed to break if struck with a force of (at least) 600 N, and 500<600.

Note that, given the partial *explanans* “500<600”, this explanation is partly an arithmetical explanation. The disposition to break if struck with (at least) 600 N may be considered to be a partially determinate property; unlike common cases of quantitative properties along the determinable/determinate divide, however, this disposition admits of non-mutually exclusive specifications; e.g. an object with mass of (at least) 60 kg can only have one maximally determinate property of mass; whereas an object disposed to break if struck with (at least) 600 N, is disposed to break if struck with (exactly) 600 N, disposed to break if struck with (exactly) 601 N, disposed to break if struck with (exactly) 602 N, etc... (Vetter, 2015, p. 55). Through the help of partially or fully determinate dispositional ascriptions such as the ones above, we may also account for comparative dispositional ascriptions such as “gold is more malleable than lead” or graded ascriptions such as “gold is highly malleable”; we may also expect such explanations to be partly arithmetical in nature.

One may wonder about the joint-carvingness of partially determinate dispositions such as “is disposed to break if struck with (at least) 600 N” *vis-à-vis* its perfectly determinate counterpart. We could, of course, borrow from old arguments against determinable properties to conclude that “is disposed to break if struck with (at least) 600 N” is not joint-carving; perhaps because there are no genuine objective similarities between different things that are disposed to break if struck with (at least) 600 N.[2] But the weight of these arguments is unclear to me; they often rely on controversial metaphysical assumptions, such as the assumption that reality is maximally determinate. The argument from similarity, as above, strikes me as particularly question-begging; it argues that determinables are not genuine properties, as objects having the same determinable might still fail to exactly resemble in determinate respects; yet one might wonder why *determinable* properties must be evaluated based on *determinate* resemblances, if not by virtue of the question-begging assumption that reality itself is maximally determinate. More simply: two red things, one crimson and one magenta, may fail to exactly resemble each other,

[2]For an overview of these arguments, see Wilson (2021), section 3.2.

but they still resemble each other to the extent that they are both red – to claim otherwise requires having prior grasp on which properties are genuine and thus establish genuine similarities, which is what we are trying to determine in the first place. The same goes for the joint-carvingness of "is disposed to break if struck with (at least) 600 N". Furthermore, the considerations against the joint-carvingness of covert dispositional ascriptions do not cut any ice against partly determinable predicates such as "is disposed to break if struck with (at least) 600 N". Once again, being disposed to break if struck with (at least) 600 N is *not* context-sensitive, at least not in the way being fragile is.

Finally, there are considerations in favour of the reality of determinable properties, such as the fact that they are needed to properly account for laws of nature, as in Armstrong (1996), French (2014), and Vetter (2015).[3] Remember that the availability of bridge-principles, in terms of simple biconditionals, does not settle the matter; to connect determinate and determinable dispositional ascriptions such as the ones above is easy enough, but this procedure does not settle *per se* a genuinely explanatory direction, and thus is not sufficient to decide matters of perspicuity.

I will not settle this matter for the time being; the crucial point is that overt dispositional ascriptions, with partially or fully determinate stimulus and/or manifestation specifications, are needed to account for ordinary covert ascriptions such as ascriptions of fragility, malleability, solubility, etc.. . . That kind of context-sensitive and vague ideology is most likely not revealing any portion of metaphysical structure.

6.2.2 *Dispositional Reality and Pluralism*

Many questions remain regarding what kind of dispositional ascriptions should be a candidate for perspicuity. On the present approach we cash out matters of perspicuity through the application of the joint-carving $\mathbb{S}$ operator to sub-sentential expressions within such ascriptions, which means that this choice turns into a choice regarding the logical form of dispositional locutions: should there be a stimulus and manifestation specification, or more than one? Or should there only be a manifestation specification, as some claim? And how should such specifications be expressed, through predicates or whole sentences? On the present approach these questions acquire a paramount importance, as it is by virtue of hypothesis concerning the joint-carving nature of these grammatical items that we purportedly acquire knowledge about metaphysical structure; thus, there is a massive payoff for our choices concerning the logic of dispositional ideology, viz. choices of which predicates, concepts, or verbs, we use to perspicuously express dispositionality. Simply put, the

[3] I will come back to this issue when dealing with the relation between dispositionality and nomicity, in Chap. 7.

present approach to metaphysical structure and realism about dispositions is a way to immediately perceive the metaphysical fallout of issues of logical form.

In a way, this is the contrary approach to Vetter's (2015, p. 141) "path of least formal resistance", in which Vetter simply assumes that potentiality displays the right logic for the job she requires, and it is only in a second moment that the philosopher should go on to explore the metaphysical import of the correspondent potentiality ascriptions: "[m]etaphysically, this is by no means the path of least resistance" (Vetter, 2015, p. 161). Things are different on my approach: if the bridge-principles provided by Vetter linking possibilities with potentialities ought to be read not as mere biconditionals, but as providing genuine metaphysical analysis (e.g., an explanation in the present understanding of the word), then this entails that the ideology of potentiality is joint-carving, and that potentiality, *as formally expressed by Vetter*, is part of metaphysical structure. That potentiality employs a somewhat artificial grammatical device then becomes a non-problem for the metaphysician, who shouldn't necessarily be tasked with the job of translating back potentiality ascriptions to more "intuitive" or "ordinary" dispositional sentences. Of course, such a level of success cannot be established without a more comprehensive evaluation of the theory vis-à-vis its competitors – which is explicitly recognized by Vetter herself (2015, p. 32).

So, how should we formulate and choose our preferred dispositional locution, viz. our candidate for dispositionality? My answer on this topic is: we shouldn't choose, for now. I don't think it would be wise to make a decision before its time, viz. to focus on certain dispositional locutions as opposed to others.

In this, I agree with Vetter: the metaphysician can help themselves with whichever formal devices they wish (if there isn't an explicit argument against them) in trying to come up with a tentative dispositional *explanans* for other phenomena in the nomic or modal family; in a second moment, the theoretical virtues of the ensuing theory will corroborate the ideological choices made as perspicuously representing the structure of reality.

We can now see an interesting issue arise, viz. how the question concerning the extent of dispositionality may turn into a question concerning the homogeneity of Dispositional Reality. Here is a question, whose complete resolution vastly outscopes this book: what if different *explananda* are better treated by irreducibly different dispositional expressions? What if, say, a certain kind of dispositional conceptual machinery is better suited to explain laws of nature, and an irreducibly different one is best suited to explain possibilities, or subjunctive conditionals? *Prima facie*, that would seem like an undesirable result: after all, there are clear connections between most members of the nomic family, e.g. between laws of nature and subjunctive conditionals; so, if both of them are to be explained through dispositions, one could think that a single dispositional notion ought to be sufficient to recapture them all. But this cannot be ensured *a priori*; if the opposite is true, and different kinds of dispositional expressions are needed to better account for different *explananda*, we could be led to embrace genuine Dispositional Pluralism, understood as the claim that different kinds of dispositional expressions are joint-carving. If Dispositional Pluralism is true, two kinds of dispositional talk with different logical form may both be perspicuous, and thus the dispositional portion of metaphysical structure – viz., Dispositional Reality– is not homogeneous.[4] Suppose that, as an instance of Dispositional Pluralism, $\mathbb{S}$(...is disposed to... if...) and $\mathbb{S}$(...is liable to...); now also suppose that:

2) a is disposed to M1 if S1
3) b is disposed to M2 if S2
4) a is liable to M3
5) b is liable to M_4

This means that, when a is disposed to M_1 if S_1, is like when b is disposed to M_2 if S_2, and when a is liable to M_3, is like when b is liable to M_4. No individual disposition or liability is reoccurring in these similarity patterns; rather, it is dispositionality and liability itself which are reoccurring. Assuming that liability is not explainable in terms of dispositionality (so understood) this instance of Dispositional Pluralism amounts to the claim that there's more than one way to perspicuously cut reality along dispositional lines. And the same goes for the other locutions in the dispositional family.

It is worth pointing out that the possibility of Dispositional Pluralism unavoidably leads to methodological complications, in the sense that it makes the evaluation of disposition-based explanations slightly more complicated. Firstly, because simplicity and number of primitives are relevant factors when choosing theories, a dispositional theory in which, say, phenomenon A is explained by dispositional primitive D_1, phenomenon B is explained by dispositional primitive D_2, and phenomenon C is explained by dispositional primitive D_3, is *ceteris paribus* worst off a theory in which A, B, and C are all explained by dispositional primitive D_1; perhaps this is but an instance of Ockham's Razor: do not postulate two things when one would do. At any rate, this makes the case for the reality of D_1 at the expenses of D_2 and D_3 (and thus, monism instead of pluralism) that much stronger. But let's remember: only *ceteris paribus*; e.g. perhaps there is an important downside in treating B through D_1, which must of course be taken into consideration. This entails not only that the evaluation of D_1, D_2, and D_3 must, to some extent, be operated independently (if they are really irreducible forms of dispositionality), but also that one must compare unified dispositional explanations with collections of distinct dispositional explanations (e.g., one in term of D_1 vs. one in term of D_1, D_2, and D_3): disposition-based theories are not only pitted against non-dispositional alternatives, but also amongst themselves.

[4] Of course, we must also keep in mind that some dispositions may be explained by other dispositions, to some extent decreasing some of this complexity. The case of determinate vs. determinable dispositions may be an example of this.

So, how are we to proceed? The crucial part is, again, that it wouldn't be wise to opt against pluralism for the get-go (which kind of dispositions should we prefer, incidentally? There's hardly a reason for choosing, at this stage of our investigation). We should proceed and select the best dispositional resources at our disposal, whichever they might be. If pluralism ends up being unavoidable, so be it; but we should always keep in mind that *ceteris paribus*, non-pluralist theories ought to be preferred. To summarize, whether we should be monists or pluralists concerning dispositionality, in the way sketched above, is an open question that cannot be settled at this stage of our investigation; we will come back to it at the end of the book, with a partial answer and a direction for future investigation: how much internal variety is there within Dispositional Reality?

6.2.3 McKitrick's Pluralism

In recent literature, pluralism about dispositions is discussed extensively at least once, in McKitrick (2018). There are important differences between our accounts, however. It seems to me that my disagreement with McKitrick lies not with our stance on Dispositional Pluralism itself, nor any other specific claim in power metaphysics, but rather with our metaphysical background.

On my account, the distinction between merely true and metaphysically perspicuous dispositional ascriptions underlies the often-drawn distinction between metaphysically innocent vs. loaded dispositional talk; commenting on Bird's (2016) similar distinction between predicatory and ontic properties/dispositions McKitrick (2018, p. 67) implicitly criticizes this distinction; for her, an abundantist perspective is preferable, as she considers the alternative to be seriously problematic when it comes to predication and similarity. Abundant properties are loosely framed by McKitrick (2018, p. 52) according to her "agnostic disjunctivism", according to which the entities playing this property-role regarding predication and similarity can be "universals, sets of tropes, sets of objects, or perhaps some other entities". McKitrick does not insist on the risky claim that there's a property for each predicate, but rather than "there are properties for the majority of the predicates we use, the kinds of things we identify, the categorizations, groupings, and distinction we make, as well as many properties that we do not care about". I will not put this loose qualification under scrutiny.

My relationship with McKitrick's abundantism, and with abundantism in general, is complicated. There's a clear sense in which the account presented in this book is abundantist. In this chapter I am using the term "property" to stand for any true ascription, and in Chap. 4 I flirted, although not without fully committing to it, to a sort of Ostrich Nominalism according to which, say:

6) a has the property of being disposed to M if S because a is disposed to M if S

Given basic disquotation principles this feels quite abundantist, perhaps even more so than McKitrick's account: firstly, there is indeed a property for each predicate truly ascripted; and secondly, the account is even more metaphysically deflated than McKitrick's agnostic disjunctivism.

That said, there is a sense in which my position is sparsist, if not in the letter, at least in the spirit. This is because for many matters of realism, say the reality of dispositions, the debate is not simply exhausted by observing true predication, and not even "categorizations, groupings, and distinction we make", to repeat McKitrick's words. These are matters not of truth but perspicuity, and thus they cannot be trivially settled without serious theoretical enquiry. The peculiarity of my account, unlike, say Lewis (1983), is that sparsism is not achieved by way of ontological inflation. Ultimately, this is where my disagreement lies with McKitrick's Dispositional Pluralism. While for her Dispositional Pluralism merely stems from the observation that the world of appearance is teeming with a multiplicity of dispositional concepts – the observation which opened Chap. 4 of this very book–, on the present account Dispositional Pluralism is a laborious metaphysical result, and for the time being, an open question.

To sum up, I do not feel misrepresented in the sparsist camp, together with Bird (2016), and all who want to distinguish metaphysically relevant uses of dispositional ideology from others, less harmless uses. But one should always keep in mind the caveat above – viz. there is an abundantist component to my position. Relatedly, it is worth pointing out that I do not find McKitrick's arguments for abundantism very compelling. There's at least two of them. The first argument is that sparsists lack an account of true predication, or, more precisely (McKitrick, 2018, p. 62):

> when a predicate is truly ascribed to some object, it is reasonable to suppose that object instantiates a property that is denoted by that predicate, *lacking an alternative account* of how this predicate ascription can be true consistent with the non-existence of a property denoted by the predicate. For the vast majority of true predications, sparse theorists lack an alternative account. By "an alternative account", I am not demanding semantic reduction of the predicate, or necessary and sufficient conditions for its application, but I would like some sort of explanation.

Based on the content of previous chapters, the reader will already see how I strongly disagree with McKitrick's first claim in the passage. Even if one accepted a correspondence-based version of perspicuity according to which sub-sentential expressions in a true predicate ascription correspond to or denote entities (as I do not: predicates do *not* denote), it would then be a further substantial metaphysical assumption to claim that all true predication is metaphysically perspicuous. I do not accept this assumption: some mere truths are not perspicuous. McKitrick (2018, p. 63) also claims that (abundant) properties are needed as semantic values for predicates. As true as it is, I am unsure how this factors in the sparse vs. abundant debate: standard Tarskian semantics deals with abundant properties-as-sets, but, again, there's little to no reason to regard semantic ascent *per se* as an increase in perspicuity. It is not the job of truth to sanction successful correspondence for all sub-sentential expressions, nor to delineate metaphysical structure. As hinted in previous chapters, I have some sympathy for deflationism about truth, according to which the job of truth is ultimately to sanction disquotation, to allow the injection of the content of the sentence into object-level discourse: if that were the case, McKitrick's demands for an explanation is bound to be unsatisfied. McKitrick asks: why is "water is wet" true? Because water is wet.

McKitrick's second argument is that sparsists make it difficult for properties to account for similarities in terms of property-sharing. The idea is that since the identification of sparse properties is a laborious theoretical enterprise, such properties do not neatly line up with the ordinary similarity judgments and categorizations that they are supposed to account for: after all, if it's hard to discover which are the sparse properties, it should also be hard to discover genuine patterns of similarities between objects. But it isn't: most crucially, we do not need to wait for the physicist, or perhaps the metaphysician, to tell us which objects resemble one another. So, if properties are needed to account for those similarities, we also shouldn't wait for an external investigation to tell us which are the sparse properties.

The sparsist's natural reply would be to claim that although we do possess the pre-theoretical ability to spot certain similarities in our surroundings, it is still a non-trivial enterprise to discover which property is responsible for that similarity – in my terms: we are in a position to pre-theoretically recognize certain similarities, although, at that stage, we do not know which piece of conceptual machinery is best apt at recapturing and expressing such similarities. The abundantist counter-objection is that this turns similarity into a mysterious matter: "sparse theorists suggest that genuinely similar things all have something, we know not what, in virtue of which they are similar. It is as if properties are invisible 'similarity pellets'" (McKitrick, 2018, p. 66).

Is that so strange? As for me, I am ready to bite the bullet. As a matter of fact, we can pre-theoretically spot patterns of similarity without being able to tell which properties are responsible for that similarity. It is not an absolute form of humility, however: on the contrary, it invites proactive investigation by rejecting the trivial answers that abundantism would produce. E.g., we see two red things. We recognize that they are similar. Why is that? It would be all too easy, and not very interesting, to simply answer that they are similar because they both instantiate the property redness. That merely amounts to labelling the problem, instead of solving it. Optics and theory of colours may have something more interesting in store for us, and the relevant conceptual machinery may have better credentials as being joint-carving. This is our path to sparse properties, or, in our terms, metaphysical structure.

In conclusion, unlike McKitrick's Dispositional Pluralism, our sparsist-in-spirit approach, in which only some true dispositional ascriptions are perspicuous, makes it unclear whether the overabundant and multifaceted nature of dispositional appearances qualifies as genuine Dispositional Pluralism. That remains to be seen.

6.3 Dispositions and Properties

There is an important lesson to learn from the previous section: some objective similarities may be expressed in a better or worst fashion, depending on the choice of predicates. E.g., we may express a certain similarity between objects a and b by saying that they are both P or by saying that they are both Q, but if $\mathbb{S}(P)$ and $\neg\mathbb{S}(Q)$ (such that, e.g. a is Q because a is P), then talking in terms of P-objects and P-similarities is an inherently better way to express matters of metaphysical structure. The flip side is that the use of certain true predications (e.g. "a is Q") may be backed by some underlying metaphysical structure that is not immediately detectable though mere conceptual analysis: a non-trivial effort is required, such as the scientific investigation that leads from "is red" to the predicates of optics.

When it comes to Dispositional Reality, this suggests the following question: how much of the ordinary or scientific non-dispositional predication is underlined by Dispositional Reality? This is the second issue that I will discuss in this chapter. It is the distillate, within the context of my account, of the old debate between Dispositional Monists, Categorical Monists and Dualists. I have introduced these traditional positions in Chap. 4: to summarize, when faced with the question "are all (sparse) properties dispositional?" Dispositional Monists answer "yes"; Dualists answer "no", but clarify that some of them are. Categorical Monists answer "no", because no (sparse) property is dispositional. A supporter of Dispositional Reality who also espouses Dispositional Monism believes that all non-dispositional ascriptions can ultimately and advantageously be explained in dispositional terms; a Categorical Monist believes the contrary (and thus claims that no dispositional ascription is perspicuous – there is no Dispositional Reality in the first place); while a Dualist rejects both claims: there is Dispositional Reality, but there is a predicative aspect of reality which is not more perspicuously expressed through dispositional ideology.[5] On the present account we shouldn't conceive the disagreement amongst these positions as a disagreement concerning how many, and which, kinds of properties there are: it is rather a disagreement concerning the perspicuity of certain ascriptions: the term "property" is used here in an abundant and deflated sense.

This means, incidentally, there is no quest to undertake concerning what makes a property dispositional or categorical; the dispositional/categorical divide is not a metaphysical divide concerning some metaphysical condition, the "mark of the dispositional" or "the mark of the categorical" partitioning the domain of properties into the dispositional and the categorical. In fact, I do not engage with properties-qua-entities being either dispositional or categorical, according to any such condition. The only distinction I require is a distinction between predicates. This is not to say, of course, that the distinction between the dispositional and the non-dispositional is merely conceptual, or metaphysically irrelevant. There is a

[5] "Monism" here should be understood as opposed to Dualism, rather than Dispositional Pluralism (as above). The labelling is a bit confusing, but unfortunately this is the entrenched terminology.

residual issue here: an issue of perspicuity. On this approach, Dispositional Monism becomes the claim that all portions of metaphysical structure revealed by joint-carving predicates are Dispositional Reality. And so, for every predicate P such that for some object a, if "a is P" is true, then there is an object b and a dispositional predicate D such that a is P because b is D (b may be identical to a, or not). The unrestricted quantification over Ps is what makes this simple version of Dispositional Monism such a radical position. If all properties must be accounted dispositionally, there are bound to be recalcitrant, almost untreatable cases: properties and relations of spatiotemporal location are famously problematic cases, as are structural or geometrical properties[6] but they are far from being the only ones. What about self-identity, or logico-mathematical properties? What is more likely, I believe, is that for Dispositional Monism to be a tenable position, some restriction must be imposed on the quantification over predicates, although it is not exactly clear how.

In what follows I will not pursue this general project, and thus I will not evaluate Dispositional Monism in its generality; I will rather discuss more specific case, viz. the prospects of giving a dispositional explanation of many ordinary and scientific properties (which is, I suspect, the kind of properties supporters of Dispositional Monism usually think about when they discuss their opinion). With "dispositional explanation" I mean the kind of explanation used so far, viz. a completion with a dispositional predicate of: x is P because... for some object x and predicate P.

6.3.1 Dispositional Accounts of (Certain) Properties

Many properties can easily be accounted in dispositional terms, both in our everyday life and in scientific contexts; as above, the ascription of some of them may require contextual factors to be fully explained. I am talking about properties such as fragility, malleability, elasticity, solubility, deadliness, flammability, irascibility, and many more.

Other properties, unfortunately, are a bit trickier to treat. Paradigmatically, this is the case of properties and relations of spatiotemporal location, and (micro)structural and geometrical properties. This is not to say, mind you, that it is impossible to come up with dispositions to pair such properties. Following in the footsteps of Mellor (1974), we may associate triangularity with the subjunctive conditional: if one were to (correctly) count the corners, the result would be three; from this conditional we may extract stimulus and manifestation specification to formulate a disposition to associate to triangularity. And the same goes for many other properties; to use another example from geometry, first popularized in Heil (2003), we may associate the property of sphericity with the disposition to roll on an inclined surface. So the idea is that we may provide explanations of such properties in terms of such dispositions.

---

[6]For a discussion of these cases, see Molnar (2003), chapter 10; also Bird (2009).

There are two broad and interconnected issues with this proposal. The first issue, I will call the Unifying Problem. Consider again the case of sphericity and the disposition to roll on an inclined surface. The manifestation of this disposition requires many other environmental factors or dispositional partners to occur. Spherical objects roll on an inclined surface if the surface is impenetrable, if the medium is penetrable, and if there is a certain gravitational attraction, with a certain orientation. Sometimes these partners are surreptitiously described in the stimulus condition, and occasionally the so-called "stimulus-manifestation" model is contrasted with a "dispositional partner" model; I believe that the difference between these two models cuts little metaphysical ice, but there is no space here to articulate an argument. What is important, is to point out that properties like sphericity confer many dispositions to their bearer, with different manifestations in different contexts, or with different partners. Spherical objects with a density similar to water are disposed to sink in water. Translating spherical objects are more disposed to generate lift when they are also rotating. And so on an so forth. Perhaps we should treat sphericity as a *multi-track* disposition, viz. a disposition with more than one pair of stimulus-manifestation specification. So we may have something along those lines:

$$\text{a is P because} \begin{cases} \text{a is disposed to } M_1 \text{ if } S_1 \\ \text{a is disposed to } M_2 \text{ if } S_2 \\ \text{a is disposed to } M_3 \text{ if } S_3 \\ \quad\quad \textit{etc}\ldots \end{cases}$$

Thus, any explanation of a property like sphericity in term of a single-track disposition is a partial explanation. The first hurdle of this proposal is how exactly the notion of a multi-track disposition ought to be understood: keep in mind that we are using "property" and "disposition" in a metaphysically deflated sense, and similarly for the notion of explanation: we are accounting for properties through dispositions only in the sense that we are providing completions of "…because…" sentences with non-dispositional predicates on the one side and dispositional predicates on the other. The simplest way to achieve this result is to understand the notion of multi-track disposition as conjunctive in character; e.g. the instantiation of sphericity might be explained conjunctively as the instantiation of several dispositions. In what follows, I will not problematize the fact that the resulting disposition might be infinitely multi-track, viz. that sphericity might only be accounted dispositionally as the conjunctive instantiation of infinite dispositions. The Unifying Problem is another. In simple words, the problem is that on this account, the association between all those different dispositions with sphericity is a primitive affair. Why do these dispositions of a spherical objects come together? This is an issue of co-variance of property ascriptions. We may also think of this way (Kistler, 2020, p. 184–185): why is it that, on this multi-track disposition account of properties, it is only those specific dispositions that conjunctively account for a property?

And why is it sphericity? And why couldn't the same collection of properties, minus one, correspond to another property?

On the present account, there is no easy answer – but there could be if we inverted the direction of explanation. This is an interesting proposal because, it seems to me, there are several connections between all the various dispositions conferred by geometrical or structural properties like sphericity and triangularity, and some of these explanatory connections relate to the original property as an *explanans*. E.g., that an object rolls down an inclined surface might be explained by virtue of its being spherical, rather than the other way round; and similarly for the other dispositions (we have not yet clarified what kind of explanation we are dealing with here, more on that below. But there is, undoubtably, an explanation here). In short, sphericity could be an explanatorily unifying factor bundling together all the dispositions associated with it; this unifying element is lost in the original direction of explanation, where sphericity is explained through the dispositions.

This suggests, in fact, another reason to prefer the other direction as genuinely explanatory. It's a fairly simple one, but whose ramifications are profound. The point, succinctly put, is that for properties like sphericity and triangularity, there is a perfectly viable definition, one which is not dispositional, but geometrical, as noted in Lowe (2010, p. 10), and Bird (2009, p. 223). A triangle is simply any polygon with three sides, or three angles; a little more elaborately, as in Wadell (1935), something is a sphere if and only if, given its volume v and its surface a, $\frac{\pi^{\frac{1}{3}}(6v)^{\frac{2}{3}}}{a} = 1$. Incidentally, it is with the help of this definition that sphericity explains the possession of dispositions such as the disposition to roll on an inclined surface. Spheres roll more easily down an inclined surface because they have a lower moment of inertia, something which depends on both their mass and geometry.[7] In fact, it is this very non-dispositional account of sphericity that allows the property to be a unifying factor for the possession of all those disparate dispositions. To invert the direction of explanation, thereby explaining the sphericity through the disposition, risks taking things horribly wrong for no clear reason.

Thus we reach the second issue. This is what I will call the Misidentification Problem; it is perhaps a vague concern, but a persistent one; it was perhaps first expressed in Mumford (1998, p. 78–81) protesting the aforementioned dispositional account of triangularity, or more precisely protesting the claim that triangularity is a dispositional property because it entails the subjunctive conditional "if one were to (correctly) count the corners, the result would be three". Mumford's idea is that

[7]Please note that the disposition to roll down an inclined surface is not exclusive to spheres: cylindrical rings are also disposed to roll, as are many other figures; this means, incidentally, that such a disposition would not be sufficient to account for sphericity, nor to be identical with it, as in Heil's (2003) original intention. We should associate sphericity with a disposition to roll down an inclined surface more easily, or faster, than any other solid.

dispositional ascriptions entail subjunctive conditionals out of conceptual necessity, whereas the same is not true for, say, the ascription of triangularity and the conditional "if one were to (correctly) count the corners, the result would be three"; translated to the current issue that would mean that while ascriptions of, say, fragility, are associated with dispositional ascriptions out of conceptual necessity, the same is not true for, say, sphericity and the disposition to more easily roll down an inclined surface: this latter association "derives from an a posteriori discovery" (Mumford, 1998, p. 78). In my opinion Mumford is right, although in the context of this book the problem is not one of conceptual analysis, but a properly metaphysical problem.

Let us articulate the difficulty more clearly. The kind of (presumably non-causal) explanation first presented in Chap. 3, and used in this very chapter, is what has sometimes called *constitutive explanation*. Remember the instances of explanation presented so far, viz. the explanation of bled and rue in terms of red and blue (or *vice versa*); the explanation of the ball instantiating redness in terms of the ball being red; the explanation of the glass' fragility in terms of its crystalline microstructure; borrowing from Dasgupta (2017) we have countless other examples of constitutive explanation: water being hot is explained by its high mean kinetic energy; losing in chess is explained by having one's king in checkmate; and so and so forth. It is not easy to articulate the constitutive aspect of this kind of non-causal explanation, and it is especially hard given the operationalist notion of explanation used so far. Traditionally, it is presented by indicating that metaphysical grounding, as an explanatory relation between facts, entails a reduction of the facts being grounded (Rosen, 2010). There's some uncertainty as to how fact-reduction should be understood, the simplest option simply being fact-identity; as noted in Audi (2012, p. 110), this simple move threatens to short-circuit the notion of reduction-entailing-grounding, given a standard understanding of grounding, as explanatory, as a strict partial order, and thus irreflexive; the logical form of grounding can at this point be questioned (Jenkins, 2011), otherwise one can reject the notion of reduction as fact-identity and enter the more elusive realm of "nothing-over-and-aboveness", or "fundamentality commitments", or "what God had to create when creating the world".

We can cash this though a just-is sentential operator of generalized identity. We have already introduced this notion earlier in the book (when discussing the Identity Theory, at the end of Chap. 4), but it is now time to articulate some details. A genuinely constitutive explanation like the one discussed so far is one of the form "A because B", such that A and B, so to speak, correspond to the same piece of reality, which is more perspicuously expressed by B rather than A. Given the deflationary account of facts employed so far, the idea of sentences corresponding to pieces of reality should not be taken at face value. Generalized identity can help us circumvent this difficulty: perhaps this kind of constitutive explanation may be recaptured as an instance of generalized identity with an explanatory reading going in only one direction. When we claim that, say, the ball being coloured is explained by the ball being red, this should be factored in the following pair of sentences:

7) For the ball to be coloured *just is* for the ball to be red
8) The ball is coloured *because* the ball is red

As noted in Linnebo (2014, p. 478–479) this pairing of just-is and because sentences would surely be repellent by Rayo's standards, according to which the very job of just-is sentences is to close explanatory gaps (Rayo, 2013, p. 54–56). But things are different here. Keep in mind that although, like Rayo, I reject Metaphsycalism^{S}, I do preserve the idea that there is metaphysical structure. A sentence need not be metaphysically perspicuous to be true, and two sentences may be flanked by a just-is operator even if one of them displays metaphysically more privileged ideology. An operator of generalized identity does not state that the two sentences have the same semantic content; in fact, it doesn't even mention the two sentences (although it uses them). More to the point, "[generalized identity between sentences p and q] also allows for representational differences between p and q", not simply in their syntax, but also "in their semantic values or cognitive uptake (e.g. the 'guise' their contents are presented under)" (Correia & Skiles, 2019). It is exactly these representational differences which allow for an explanation, in the primarily epistemic sense discussed in Chap. 3, to be compatible with generalized identity: such an epistemic context is surely opaque enough to bar substitutionality as sanctioned by generalized identity.

The methodology laid out so far submits that an epistemically successful explanation, embedded in a more virtuous theory, has good credentials of expressing an increase in representational fidelity from *explanandum* to *explanans*, which we reconstructed through the notions of joint-carvingness and perspicuity. Within the background of metaphysical structure and perspicuity employed so far, the truth of sentence pairs like (7)–(8) amount to the fact that there are metaphysically more and less privileged ways to talk truly about the same portions of metaphysical structure.

This is where we encounter problems concerning the explanation of properties in terms of dispositions. Although the two predications are connected, there is some resistance to the idea that the worldly content of "the ball is spherical" is more perspicuously expressed by saying "the ball is more easily disposed to roll". That is to say, the following pair is much more controversial:

9) For the ball to be spherical *just is* for the ball to be more easily disposed to roll
10) The ball is spherical *because* the ball is more easily disposed to roll

Why is that? Nevermind the fact that if generalized identity is as transitive and symmetric as standard identity (Correia & Skiles, 2019, p. 4) then the multiplicity of dispositions associated with sphericity, as above, is enough to make (9) false, given that such dispositions are all patently distinct from each other. But even correcting this account in a multi-track fashion there is a lingering feeling that something is wrong. Sphericity has a very clear geometrical and thus non-dispositional definition, thus it would be very natural to think that sphericity and the disposition to more easily roll down an inclined surface are different portions of metaphysical structure.

This is not, mind you, a knock-down argument; on the contrary, it is based on a very resistible intuition. But once we take a step back we realize that the rejection of (10) allows us to solve both problems more easily. To summarize, sentence pair (9)–(10) expresses a dispositional account of sphericity; the account has two components: the explicitly explanatory claim (10), and the generalized identity claim (9). The Unifying Problem threatens (10), by suggesting that it might have the direction of explanation wrong. The Misidentification Problem threatens the generalized identity claim (9), by putting forward a metaphysical version of Mumford's conceptual worry that for the ball to be spherical is not for the ball to be more easily disposed to roll. That sphericity is not to be identified with a disposition, and is characterized in a different way, is what allows us to construct an explanation of the disposition through the property; e.g. spheres roll more easily down an inclined surface because they have a lower moment of inertia, which in turn depends on their geometry, and thus sphericity.

It should be noted that geometrical properties like sphericity are not the only ones for which the Unifying Problem and the Misidentification Problem arise. In fact, as recently discussed in Kistler (2020), similar issues arise concerning the dispositional treatment of certain properties of fundamental particles, e.g. electric charge. This is particularly troubling, since properties like electric charge seemed like the perfect candidate for a dispositional treatment (remember the discussion on the Argument from Science from the previous chapter!). The Unifying Problem is perhaps more troublesome here. Philosophers like to associate electric charge with a disposition to attract or repel particles with a magnitude calculated with Coulomb's Law, or perhaps as a disposition to induce an electromagnetic field. Therefore, a dispositional explanation of electric charge might be formulated by providing the following pair of sentences:

11) for particle p to have electric charge q_1 *just is* for particle p to be disposed to attract particles with charge q_2 at distance r with force $k_e \frac{q_1 q_2}{r^2}$
12) particle p has electric charge q_1 because particle p is disposed to attract particles with charge q_2 at distance r with force $k_e \frac{q_1 q_2}{r^2}$

Yet this disposition is not the only one associated with electric charge. Kistler (2020, p. 176) offers a non-exhaustive list of dispositions associated with electric charge:

> The disposition of x to undergo a force $\vec{F} = q\vec{E}$ if x is in electric field $\vec{E}$.
> The disposition of x to attract a second object with charge q* and at distance r with force $F = k_e \frac{q \cdot q^*}{r^2}$ (Coulomb force).
> The disposition of x to undergo a force $\vec{F} = q\vec{v} \times \vec{B}$ (Lorentz force) if x is moving with $\vec{v}$ in $\vec{B}$.
> The disposition of x to bear a magnetic moment $\vec{\mu} = ½\vec{r} \times \vec{v}$ if x is in rotation with speed $\vec{v}$ on radius $\vec{r}$.

At this point, we find ourselves in a familiar pickle. A multi-track treatment of electric charge as a disposition is desirable; in the simplest form, this amounts to a conjunctive treatment in which an ascription of electric charge is explained by a conjunction of dispositions such as the ones above. But given this multi-track treatment, the Unifying Problem and the Misidentification Problem arise again. How to solve them? It seems to me that (at least) two solutions are available here.

6.3.2 First Solution: Explanatory Inversion

As suggested before, the easiest solution here might be to invert the order of explanation in (10), (12), and the like. Sphericity is not explained by a disposition to roll, it's the other way round; electric charge is not explained by a disposition to attract or repel, it's the other way round; and so on. On this account, charge is not to be identified with any such disposition (solving the Misidentification Problem), and rather functions as an explanatorily unifying element that explains why they always come together (solving the Unifying Problem). In Kistler's (2020, p. 177) words, "[t] he reason to postulate the existence of electrical charge q is that the hypothesis of the existence of this intrinsic property provides the best unifying explanation of why the different dispositions [. . .] mentioned above always come together".

This explanatory inversion is common amongst many grounding theories of powers, as the useful taxonomy in Tugby (2021) demonstrates; viz., it is a common move to claim that property instances such as instances of electric charge fulfil the task of grounding/explaining/making-true the dispositions/dispositional ascriptions of the propertied entity.

As popular as it is, however, this first solution strikes me as very problematic. The problem is still one of identification: on the present account, the property of charge is not to be identified with any of the different dispositional tracks usually associated with it. But what is electric charge then, that warrants the kind of explanatory roles assigned to it? No solution seems satisfactory. We can illustrate this difficulty in the form of a two-horn dilemma. The two horns are generated by the two possible answers to the question: has the property doing the explaining (e.g. sphericity, electric charge) a dispositional nature, or not?

Let us begin with the second horn: the property doing the explaining does not, strictly speaking, have a dispositional nature. This is where we find most grounding theories of powers. This stance is usually meant as a reply to the determinative/ explanatory circularities of Bird's dispositional essentialism, in which properties get their identities fixed by the dispositions which they in turn ground. So, the idea goes, we get rid of the "identity fixing" part: the property does not have a dispositional essence *per se*, and its identity is not fixed by the dispositions that are usually associated with it. Rather, the former grounds the latter. This doesn't necessarily mean that the nature of the property is a primitive quiddity about which nothing can be said: after all, sphericity has a geometrical definition.

The main problem here is that dispositionality is effectively being explained away, expunged from what would otherwise be relevant explanatory bases. In fact, provided the resulting theory to be successful enough, by the very standards set out in this book, that would mean that such dispositional ascriptions are less perspicuous than ascriptions of, say, sphericity or electric charge. A realist about dispositions cannot be too happy with this. Although not always recognized, this is a strange predicament for the grounding theories of powers: one would expect friends of dispositions to argue for dispositions to be fundamental, and not derivative. Alas, that is not so. Relatedly, the position incurs in the old categoricalist problem of always having to find a non-dispositional element to ground the dispositions. The case of sphericity, as above, might be easy enough; but what exactly is this non-dispositional, qualitative element of electric charge that explains – in a unifying fashion – all the dispositions?

One may try to escape this charge by claiming that the property doing the grounding/explaining is itself genuinely dispositional, in the sense that it satisfies a special metaphysical condition that makes it a "dispositional property" or a "power", as in Azzano (2019, 2020) and Kistler (2020, p. 177). Within the context of this theory, this condition is supposedly simply met by doing the grounding. But if all it takes for there to be powers, is for there to be full grounds for dispositions, the position is easily trivialized: after all *something* or other will ground dispositions, according to all accounts, non-Humean or not. If no modal or dispositional element is detectable in the grounding base, it is unclear to me why the grounding theory's property of electric charge warrants being called a power in the first place.

We may solve this difficulty by embracing the other horn of the dilemma, and claiming that properties like electric charge *do* have a dispositional nature after all. This is where we find your standard dispositional essentialism. Now the problem is: if electric charge has a dispositional nature, viz. a nature which can be characterized in terms of dispositions, which dispositions are these, exactly? If electric charge is to count as a dispositional property in this sense, we should be able to write it down as a disposition on the page, which means that we should be able to select stimulus and/or manifestation specifications for it. But which ones?

These dispositions simply can't be the dispositions usually associated with charge, either individually or collectively – because remember, charge is supposedly explaining and unifying them. A property cannot explain what constitutes its nature: an explanatory circularity is afoot here.[8] If, on the other hand, electric charge is to be identified with a *sui generis* disposition, different from all the dispositions above but properly explanatory and unifying with respect to them (let us call this the "ur-disposition" of electric charge), then we are still owed an answer to the question: which disposition is it? This question is hard to answer, for one specific reason: science does not discover "ur-dispositions", whatever they are. It simply discovers

[8] See Jaag's (2014) criticism to Bird's (2007a) dispositional essentialism.

dispositions. That is to say, it allows us to characterize electric charge through the scientifically-informed dispositions mentioned by Kistler – which is on the present account count as *explananda*, rather than *explanans*. So, what *other* disposition might there be to characterize electric charge, that warrants it having a dispositional nature? This strikes me as an idle postulation of metaphysics.[9]

All in all, neither horn of the dilemma is particularly appealing. On the explanatory inversion we are considering, either the *explanans* property does not have a dispositional nature – thus effectively explaining away dispositionality– or it does – thus postulating a mysterious dispositional item beyond the ones we are acquainted to. Where does that leave us?

6.3.3 Second Solution: Non-constitutive Explanation

Despite these difficulties, I believe that the explanatory inversion with respect to (10) to (12) is the right move here: it is independently motivated and convincing, in fact, it is an *actually available* explanation employing resources from geometry and physics respectively. Thus, the disposition to roll is explained by sphericity. Dispositions of attraction or repulsion are explained by electric charge, as are the other dispositions in the list above, and their mutual interrelations.

The problem of this explanatory inversion, explored in the first horn of the dilemma above, is that it threatens the correspondent dispositional talk with non-perspicuity. A supporter of Dispositional Reality need not claim that all dispositional talk is perspicuous; but it would be a little disheartening if a charged particle's dispositions to exert Coulomb forces would turn to be non-perspicuous, given the fundamentality of such dispositions and their prominence in the overall literature. If the extent of Dispositional Reality is so minimal to not even cover dispositions in fundamental contexts, then the prospects for realism about dispositions as a whole become grim.

There is a natural solution to this difficulty: deny that the non-causal explanation in question is constitutive in the first place. Based on what has been said before, we may characterize constitutive explanation "because$_C$" through the following principle (for any two sentences p and q, and formalizing the just-is operator as "≡"):

Ce) $p \text{ because}_c\ q \rightarrow p \equiv q$

---

[9]There could be a way to claim that electric charge is a disposition, albeit not identical to any of the dispositions above: it could be a disposition in Molnar's (2003, p. 194–198) sense, viz. as having, as manifestation, not any of the overtly measurable manifestations, but the contribution that the charged object is making in any such final effect. To understand what this contribution ultimately amounts to, however – both in its specificity and in its more general metaphysical categorization–, a far more radical reboot needs to be undertaken concerning the modelling of dispositions, manifestations, and causation, which I will not pursue here; although for some criticism of Molnar's three-step contribution model, see McKitrick (2010).

Please keep in mind that (Ce) is only a conditional, and not a biconditional. I am *not* defining constitutive explanation through identity, generalized or otherwise; I am merely characterizing the constitutive aspect of this kind of explanation. Even with this clarification in mind, (Ce) is only a useful approximation, for a number of reasons.[10] That said, it is informative enough for the point we are trying to make here.

[10]Two complications to keep in mind are the following. First, there is an issue of adicity. Constitutive explanation is taken to be variably polyadic: many things, taken together, can constitutively explain one – this is surely the case if we want to model constitutive explanation along the lines of Fine's (2012) logic of ground. But the operator of generalized identity standardly is not. I propose to patch this difficulty by stipulating a *variably polyadic operator of generalized identity*. The idea of variably polyadic identity may seem like heresy: how can many things be identical to one? But we are talking about *generalized* identity here: when we state that p just is q, we are stating that p and q are describing the "same features of reality" (Linnebo, 2014, p. 468); we are simply stating that in a way that does not require the reification of those features as facts to which we can refer to through nominalizations starting from "p" and "q"; so there is nothing wrong to claim that if p just is q, r, then p and q and r (the latter two taken collectively) similarly describe "the same features of reality". E.g. for Valeria to like pizza and ice-cream just is for Valeria to like pizza and Valeria to like ice-cream (p = "Valeria likes pizza and Valeria likes ice-cream", q = "Valeria likes pizza", r = "Valeria likes ice-cream"); in this sense, the behaviour of generalized identity (and indirectly, constitutive explanation) *vis-à-vis* conjunction is embedded in its very logical form. So, for our purpose, we may take both "because$_C$" and "just is" as variably polyadic – and so, (Ce) is a simplified version, as limiting case.

The second complication is a bit more elusive. But the need for it can be immediately expressed by considering that (Ce) has at least two straightforward counterexamples, assuming that because$_C$ obeys your standard logic of ground. In such a logic, both disjunction and existential quantification can receive multiple distinct full explanations. Assume that $p \bigvee q$, while both "p" and "q" are true. By the standard logic of grounding, we have that $p \bigvee q$ because$_C$ p, and $p \bigvee q$ because$_C$ q. By virtue of (Ce), $p \bigvee q \equiv p$ and $p \bigvee q \equiv q$; but given that $\equiv$ is meant to be an equivalence, we reach the unacceptable conclusion that $p \equiv q$. Something similar happens for existential quantification. What to do? I do not wish to complicate things further by rejecting the equivalence nature of generalized identity, or rejecting the standard logic of ground for constitutive explanation. I would rather claim that (Ce) requires a bit of fine-tuning.

Correia and Skiles (2019) offer a fascinating way to understand grounding in term of generalized identity through disjunctive parthood: p is a disjunctive part of q ($p \lesssim q$) if and only if, for some r, $q \equiv p \bigvee r$. Again, for our purpose we only need a conditional, instead of a biconditional. Thus, with some simplification, the conditional in question would be:

Ce*) p because$_C$ q $\rightarrow p \lesssim q$

(Ce*) avoids the problems above, capturing the idea that if q constitutively explains p, then p is but one of the ways in which q could be realized. (Ce*) might considered problematic as it dictates that all constitutive explanation is inherently disjunctive in nature, viz. that any *explanans* just is the *explanandum*, *or* some other thing; but this charge can be rebutted simply by claiming that in certain cases this "additional" r is q itself, viz. the *explanandum* (improper disjunctive parthood). See Correia (2010, p. 272) and Correia and Skiles (2019) again for this issue. In conclusion, by virtue of improper disjunctive parthood being a limiting case of disjunctive parthood, (Ce) can be taken as a limiting case of (Ce*).

Some instances of (Ce) may be the following: if I have lost this game of chess because$_c$ my king is in checkmate, then for me to have lost this game of chess just is for me to have my king in checkmate; if the glass is fragile because$_c$ it has a crystalline microstructure, then for the glass to be fragile just is for the glass to have a crystalline microstructure. Other instances of constitutive explanation presented in this book follow the same pattern. When in some instances of constitutive explanation, e.g. p because$_c$ q, the representational difference between p and q which makes one direction genuinely explanatory suggests an increase in perspicuity between p and q. This is what was problematic in the biting-the-bullet strategy in the previous section; if we invert the direction of explanation from (10) and (12), we run the risk of constitutively explaining dispositional properties in terms of non-dispositional properties, thus effectively making the correspondent dispositional talk less perspicuous.

But this needn't be the case. Not all non-causal explanation in metaphysics need be constitutive, in the sense that not all such explanations need obey the (Ce) principle. In metaphysics there is also room for non-causal explanation between fully distinct phenomena; in fact, in the next chapter I will present one of such candidate, which is usually thought to separate neo-Humeans from non-Humeans metaphysics: a phenomenon called *governance*. Without getting bogged down in details, constitutive explanation – as modally robust as it is– does not violate any principle of neo-Humeans metaphysics; the ball being red explains, and thus may very well necessitate the ball being coloured, but a neo-Humean would be hard-pressed to consider this a problematic "necessary relation between distinct existences"; this is recaptured in our account by principle (Ce), claiming that the ball being red just is the ball being coloured. However, non-constitutive explanations, such as governance, in which (Ce) fails, provide some genuine "metaphysical pushings and pullings" (Sider, 2011, p. 145) that a neo-Humean would find suspicious. As realists about dispositions, however, such things are not taboo, and we may think about deploying them if the need arises.

This is important because a non-constitutive non-causal explanation, disobeying the (Ce) principle, is not to be understood as providing in the *explanans* a more perspicuous way of describing the same portions of metaphysical structure; and similarly, there is no compulsion to think of the *explananda* as less perspicuous. The two sides are distinct phenomena, and thus the existence of an instance of explanation between them does not sanction, neither directly nor indirectly, an increase in perspicuity. Let us call this non-constitutive explanation "…because$_{n\text{-}c}$…"; thus, it may very well be the case that p because$_{n\text{-}c}$ q even though both p and q are metaphysically perspicuous. Both p and q reveal distinct portions of metaphysical structure; and yet there is some non-Humean pushing and pulling between them.

This is what is going on, I believe, between sphericity and the disposition to more easily roll down an inclined surface; as the Misidentification Problem suggests, there is something deeply unsettling in the claim that for an object to be spherical just is for the object to be more easily disposed to roll down an inclined surface (or *vice versa*), something straightforwardly entailed from a constitutive explanation between the two – in either direction. To say that something is spherical is not a metaphysically better or worst way to say that it is so-and-so disposed: sphericity and the disposition appear to be two distinct things. Yet, as the Unifying Problem suggests, there is some (non-constitutive) explanation between the two, just as there is some non-constitutive unifying explanation between electric charge and the dispositions listed by Kistler. Therefore, switching the explanatory direction in (10) and (12), and switching from a constitutive to a non-constitutive explanation allow us to solve both the Unifying Problem and the Misidentification Problem without giving up the idea that the dispositions discussed in these pages are real, or no less real than those other properties.

There's a further virtue to the theory: under the assumption that this non-constitutive explanation is a necessitating explanation, viz. it entails a strict conditional, the non-constitutive explanation from properties to dispositions entails, *per* Chap. 5, the kind of modal fixity required to eschew talk of Ramseyan humility; amongst other things, it surely entails the principle that I have dubbed (MF).

The move doesn't come for free, however. The price to pay is that, against our initial intention, there is no way to properly account for properties like triangularity, sphericity, and electric charge in a dispositional fashion. There's little we can say on the matter: sphericity is not a disposition. Electric charge is not a disposition. But they both bring real dispositions with them, in the non-constitutive and non-Humean way sketched above. E.g., it may both be a part of reality that particle *e* has electric charge q_1, and that particle *e* is disposed to... etc....; but the former non-constitutively explains the latter, which also explains why all dispositions associated with electric charge come bundled together. This is compatible with particle *e* having electric charge q_1 not being part of Dispositional Reality. To use a slogan, dispositions come prepackaged together, but the package itself is not dispositional.

6.4 The Bigger Picture, and an Example

We have said many things concerning the dispositional treatment of everyday and scientific properties. It is now time to take a step back and observe the bigger picture. We may start by considering a simple predicate like "fragility". Two things must be observed:

(a) fragility is associated not to a single disposition, but many
(b) the association between fragility and each of these dispositions is mediated by context

Concerning point (a), the multi-track nature of supposedly dispositional properties is perhaps even more widespread than initially thought, and concerns everyday properties like fragility as well. As in Vetter (2015, p. 40):

> [i]t is easy to see that most of our everyday dispositions are multi-track. A fragile glass may manifest its fragility in breaking upon being hit with a spoon, being dropped onto the floor, being sung to by a soprano, or being subjected to pressure over a period of time. Fragile parchments break upon being merely touched, and a fragile old wooden chair may split when transferred into a different temperature. Irascible people may manifest their irascibility by becoming angry upon being yelled at, being disagreed with, or being politely told to wait.

The hope of reducing these tracks to a single stimulus and/or manifestation specification are grim. On the one hand, as I have argued, the prospects for such a unification would appear to depend on a properly scientific investigation, rather than a metaphysical one; viz. it is for science to decide whether there is a single underlying disposition underpinning all fragile things, or if fragility is an irreducibly multiply realizable property. On the other, Vetter (2015, p. 41–42) has a proper argument to the conclusion that generalizing stimulus and manifestation of fragility leads to miscategorizing clearly fragile things as non-fragile. For our purpose, we may take the multi-track nature of fragility at face-value.

About point (b), we have seen in this chapter how semantically unruly ordinary predicates like fragility are; they may be constitutively explained through dispositional ascription only with the mediation of a context. Thus, putting (a) and (b) together, a property like fragility may be constitutively explained through dispositions in a very complex way, e.g. according to the following schema:

$$\text{a is fragile because}_c \begin{cases} \text{a is disposed to } M_1 \text{ if } S_1 \text{ in context } C_1 \\ \text{a is disposed to } M_2 \text{ if } S_2 \text{ in context } C_2 \\ \text{a is disposed to } M_3 \text{ if } S_3 \text{ in context } C_3 \\ \textit{etc}\ldots \end{cases}$$

As we have seen, this multi-track treatment was unavailable for properties like electric charge. So what is the difference between electric charge and fragility? This is what we now pass to discuss.

6.4.1 Disjunctive Vs. Conjunctive Multi-Track

We have seen how electric charge is (allegedly) multi-track and we have seen how fragility is multi-track. But there is a crucial distinction between them. Electric charge is conjunctively multi-track, while fragility is disjunctively multi-track. What do I mean by that? We have seen how many different dispositions can be associated with properties like sphericity and electric charge; but any object possessing such properties possess all such dispositions together – this is why the Unifying Problem arose in the first place, because anyone wanting to account for sphericity dispositionally still had to explain why all those dispositions came together. The same is not true for fragility: for an object to be fragile, it needs to possess only one of the correspondent dispositions (observe how Vetter uses "or" in the passage above when describing the different tracks of fragility). Objects possessing one disposition usually do not possess the others, hence there is no bundling of dispositions here.

The difference between conjunctively and disjunctively multi-track dispositions has mostly gone unnoticed in the literature. Molnar (2003) and Vetter (2015) think of multi-track dispositions in disjunctive terms, whereas Heil (2003) and Williams (2011b) clearly think of multi-track in conjunctive terms; Lowe (2010) does so as well, as it considers whether magnetism could be considered a disposition to attract ferrous metal, *and* a disposition to induce electrical current, *and* a disposition to produce a kind of sensation in pigeons, etc.... Clearly, this is not so different from how we have been treating electric charge in this very chapter.

This uncertainty in the literature does not help, for the difference between conjunctively and disjunctively multi-track has widespread consequences. First and foremost, while the dispositional multi-track account may work for properties like fragility, it won't work for conjunctively multi-track properties like sphericity and electric charge, for reason extensively discussed in previous sections. In those cases, we may need to provide explanations for those dispositions, thus explaining why they all bundled together: this explanation is usually provided by the property itself, thus nullifying the possibility of the explanation running in the other direction, as originally intended.

It may be worthwhile to notice that for properties like fragility the Misidentification Problem is less gripping as well. There is no resistance to the notion of constitutively explaining fragility as a disposition to break if struck, even with all the caveats and qualifications above: on the contrary, it seems like the right move to claim that for an object to be fragile just is for the object to be disposed to break if struck. While disjunctively multi-track explanation of properties in term of dispositions tend to function, the same does not happen for conjunctively multi-track accounts, in which such explanation is plagued by the Unifying Problem (which lead us to invert the direction of explanation) and by the Misidentification Problem (which lead us to shift to non-constitutive explanation). Long story short: multi-track dispositional accounts of properties function when they are disjunctive (e.g. fragility) but not when they are conjunctive (e.g. electric charge).

6.4.2 *Multi-Track Deflated*

Finally, it ought to be noticed that, on my account, it is highly questionable whether there could be genuinely multi-track dispositions in the first place. This may seem a rather bold claim, given how widespread multi-track dispositions are in the literature, but it is an unavoidable consequence of my metaphysical set-up.

Roughly put, a multi-track disposition is a single disposition which can be characterized with more than one pair of stimulus/manifestation specification. But as noted in Williams (2011b) there is a certain ambiguity when discussing multi-track dispositions: on the one hand, accepting multi-track dispositions could mean that a single property conveys multiple dispositions to its bearer; on the other hand, perhaps the most common reading of the expression, we mean that a disposition itself is characterized by more than one pair of stimulus/manifestation specifications.

The reader of Chap. 4 will perhaps remember that the distinction between "dispositions" and "properties" is one of the many lexicographically convoluted ways to express a distinction between metaphysically innocent and loaded dispositional talk. On my account, this distinction is de-ontologized and unshackled by the layered conception of the world: there are no "lightweight dispositions" and "heavy-weight dispositions". There is only *true* dispositional talk and *perspicuous* dispositional talk. Consequently, the very notion of a multi-track disposition acquires a different connotation here.

According to the first sense, a case of a multi-track disposition is simply the case of a property which confers more than one disposition to its bearer. This is a rather harmless claim, and one which I account for with my "bundling of dispositions through properties" through non-constitutive explanations seen above. But as Williams (2011b) notes, this is patently not what is meant by multi-track in general: the point is rather that of a single disposition being associated more than one specification pair. With "disposition" here we simply mean true dispositional ascriptions, the deflated sense used throughout this chapter. But on this account it is unclear how could a single disposition be multi-track: there is no unifying disposition-as-entity or property-as-entity that may be considered multi-track. There merely are various dispositional ascriptions, which may be bundled together, viz. co-varying in the same subjects of ascription; and of course the ascriptions themselves are different, if anything at the level of predicates which is relevant here. The only sense in which there could be a multi-track disposition is if we could establish some sort of identity between the various dispositional ascriptions, to recapture the idea that we are talking about the *same* disposition. This may be achieved, again, by our operator of generalized identity. E.g., we may recapture the idea that electric charge is a multi-track disposition by claiming that

1) for particle x to be disposed to undergo a force $\vec{F} = q\vec{E}$ if in electric field $\vec{E}$ *just is* for particle x to be disposed to attract a second object with charge q* and at distance r with force $F = k_e \frac{q \cdot q^*}{r^2}$

Yet sentences like (12) may strike the reader as absurd, and reasonably so. The Unifying Problem was raised exactly because the various tracks are distinct, and their bundling requires a unifying explanation. In conclusion, there is no genuinely multi-track disposition on my account; multi-track accounts of properties are simply accounts of properties treated as conjunctions or disjunctions of dispositional ascriptions, as the case might be. Results, as above, widely vary.

6.5 Conclusions

Even conceding that there is Dispositional Reality, how much of reality is dispositional? And is Dispositional Reality a homogeneous realm, or is pluralism to be preferred? I have offered a partial answer to both questions by asking which different kinds of ascriptions, both dispositional and non-dispositional, are ultimately underpinned by Dispositional Reality. A twofold conclusion has been reached.

Firstly, while some non-dispositional ascriptions can be explained dispositionally, albeit tortuously (thus suggesting that they are non-perspicuously describing some portion of Dispositional Reality), other ascriptions are resistant to this kind of treatment. The results, in this regard, are scattershot. Some properties like sphericity and electric charge, if real at all, do not belong to Dispositional Reality, although there are meaningfully connected to it by instances of non-constitutive explanation. These explanations may very well entail a sort of necessitation, in the sense that it is may be metaphysically necessary that if something is electrically charged, is disposed to behave according to Coulomb's Law. This will become crucial, as we will see in the next chapter, in accounting for laws of nature through dispositions.

Finally, it is worth pointing out again that on the present account there is no metaphysical divide between dispositional and categorical properties, but the result still qualifies as a form of Dualism: Dispositional Reality may very well be a vast region of metaphysical structure, but there is little reason to think that it exhausts it.

The second conclusion is that it is too soon to decide whether only one kind of dispositional ascriptions are perspicuous, and which one. The choice between Dispositional Pluralism and its adversary is an open question that depends on the explanatory successes of dispositionalism and the resources required to achieve it.

Chapter 7
Explaining with Dispositions

7.1 Introduction

The appeal of dispositionalism, understood as the endorsement of Dispositional Reality, vastly depends on its credentials as a body of explanations. This is another way to state, again, that the beating heart of dispositionalism lies in its applications, viz. in the successes of our efforts to explain regularities, laws, counterfactuals and other modalities through the appeal to dispositional resources. It is finally time to pursue some of these explanations.

We're not going in completely blind: in fact, we already know something about the kind of explanations we seek. Firstly, we are working with a deflated notion of explanation; viz. we are seeking for the completion of "...because..." sentences with a dispositional ascription on the side of the *explanans*; although it is arguably difficult to extrapolate genuine explanations from this primarily epistemic sense of the notion, we have seen how the postulation of such supposedly explanatory patterns can lead to genuine metaphysical progress: as in the case of the bled/rue alien tribe, the more theoretically virtuous amongst these explanations eventually make it out on top. Secondly, we should presuppose very little about the logical form of both *explanans* and *explanandum*; in the case of the dispositional *explanans*, the possibility of Dispositional Pluralism does not force us to restrict our attention to a dogmatically selected subset of canonical dispositional expressions, whichever they might be. We must keep our eyes open – but we must also remember that *ceteris paribus*, pluralism is more costly than monism. On the present approach to metaphysical structure and perspicuity, the issue of logical form is paramount: whether certain dispositional locutions or others are selected as *explanans* has wide-ranging consequences about the make-up of reality.

Thirdly, and finally, I would like to draw attention to the dispositional *explanans* themselves. I have said that such an explanatory item should take the form of a dispositional ascription to some entity, of whichever logical form. For someone well-versed on the literature, this may already strike as unduly restrictive.

L. Azzano, *Dispositional Reality*, Synthese Library 482,
https://doi.org/10.1007/978-3-031-52625-1_7

Dispositional *explanans* may take the form of simple dispositional predication, but not necessarily; e.g., the usual dispositional essentialist's explanation of regularities and laws, is not simply in terms of something being so-and-so disposed, but is usually expressed as dispositional properties having a certain essence. To employ a toy-example, soluble things dissolve in water because it is in the nature of solubility, as a dispositional property, that soluble things dissolve in water; what is really explaining, here, is the property's nature.[1]

My stance on the matter is the following: I would like to avoid such complications, for two strictly interrelated reasons. Firstly, as I will argue in the next chapter, I am very sceptical that talk of essence of dispositional properties, as widespread as it is in the literature, is genuinely explanatory. Secondly, essence is itself an ideological cost; maybe not in a strictly Quinean sense, but its use in such disposition-based explanations still incurs in a metaphysical cost. Even if we switch from an objectual to a generic essence as in Vogt (2022), we will still need a sentential operator to express essence, say $\Box_e$, for which it is meaningful to ask whether $\mathbb{S}(\Box_e)$. Any dispositional theory that displays $\Box_e$ in the *explanans*, therefore incurs in further theoretical costs: for it is not only a theory about the reality of dispositions, but also about the reality of essences. *Ceteris paribus*, I would like to keep the theory purer, and avoid such additional machinery; if essences will prove useful, I will gladly add them (although, as I said, I will claim in the next chapter that such machinery can hardly prove itself useful in the way usually espoused by friends of dispositions); at any rate, it doesn't seem wise to me to deploy essentiality willy-nilly, just to keep up with the literature. Let us start with simpler dispositional *explanans*, and see where that gets us. Furthermore, there may be inherent benefits in trying to exclude essence from our explanatory bases: as a smaller number of primitives is itself a theoretical virtue when it comes to comparing positions, within the dispositionalist camp and beyond, it may be worthwhile to develop my take on Dispositional Reality by cultivating the metaphysically deflationary flavour of the previous chapters, one that – unlike standard dispositional essentialism– does without properties and their essences: once again, *ceteris paribus* an explanatory hypothesis on a sparser basis is to be preferred. In other words: in laying the foundations for an explanatory theory concerning Dispositional Reality with some winning theoretical virtues, we are already providing partial corroboration for the view itself, viz. for the view that the simple dispositional ascriptions ultimately doing the explaining, unshackled by property-talk, instantiation-talk, and essence-talk, are metaphysically perspicuous.[2]

---

[1] E.g., in the standard dispositional essentialist's derivation of laws and regularities from dispositions, as in Bird (2007a, p. 46) a crucial premise is one concerning the essence of dispositional properties, or "potencies" as Bird calls them; it may be worth noticing that Bird's understanding of a dispositional property is in terms of counterfactuals. More generally, "[o]n the dispositional view, the explanation is provided by the essential natures of the fundamental properties" (Tugby, 2016).

[2] I will come back to essence-based explanations in Chap. 8, where my approach will be vindicated. As widespread as it is amongst friends of dispositions, essence-talk shouldn't be included amongst the primitives of our account of Dispositional Reality.

One final consideration before we get to the actual explaining. A brief survey of the book's contents will reveal that only some explanatory endeavours are attempted here; I do not attempt a disposition-based theory of causality, nor a disposition-based theory of modality or counterfactuals. These are too vast topics to be tackled here. I will instead focus on regularities and laws of nature for a simple reason: the dispositional treatment of such items will allow us to talk about the topic of *governance*, a crucial notion to delineate the non-Humean element of Dispositional Reality, and one which is very useful to position the theory within the current landscape; of course laws and non-accidental regularities are closely related to modality, counterfactuals, causation, and so much more. But there isn't space for everyone here. This limited scope of the present research naturally has some consequences. Most importantly, this means that my preferred variant of dispositionalism *qua* explanatory hypothesis can only be partially pursued, and that the comparisons it can be subjected to, within the dispositionalist camp and beyond, are similarly limited. Ultimately, this is the reason why I qualified the very beginning of this book by claiming that I can only provide partial motivation for the claim that there is Dispositional Reality. The theory is only partially presented, and thus its virtues can only partially emerge: the rest, once again, will have to wait.

In Sect. 7.2 I will introduce the star of this chapter, viz. the peculiarly non-Humean kind of explanation that is sometimes called "governance". This will lead us to a wider discussion on the different kinds of explanation required for the task at hand, and the place of governance within this explanatory zoo; this will be the topic of Sect. 7.3. With this set-up in mind, in Sect. 7.4 I will describe how to explain non-accidental regularities with dispositions, while in Sect. 7.5 it will be time to explain laws of nature with dispositions. Crucially, two different kinds of explanations (albeit interestingly related) will be required for these two jobs. Finally, in Sect. 7.6 I will offer some concluding remarks.

7.2 Meet Governance

Power metaphysics is often considered an instance of non-Humean metaphysics. The distinction between neo-Humean and non-Humeans metaphysics is often vividly perceived, albeit not always neatly drawn; traditionally, the bone of contention was the so-called Humean Supervenience: "all there is to the world is a vast mosaic of local matters of particular fact, just one little thing and then another. [. . .] All else supervenes on that" (Lewis, 1986, p. ix–x). Humean Supervenience was not initially framed as an explanatory thesis, but has been in the ensuing literature, as in Loewer (1996), and Maudlin (2007, p. 172).

At any rate, the distinction between neo-Humeans and non-Humean may be explicitly characterized as an explanatory distinction, and more specifically as a distinction concerning a peculiar kind of explanation called *governance*.[3]

The idea behind governance is intuitive enough: assuming that reality is ordered, at least in the loose sense that from the distribution of properties across space-time patterns sooner or later are bound to emerge, supporters of governance claim that the emergence of such patterns is not a brute unexplainable fact. This is something that a neo-Humean or Lewis-inspired metaphysician would reject: but for the non-Humean metaphysician reality is imbued with a special aspect – let us keep neutral about it for now– that explains and determines why things behave in the way they do with a certain regularity; opinions then diverge as to what exactly that is.

Traditionally, the simplest way to understand governance is through necessitation, as in the DTA theory of laws.[4] According to DTA, laws are second-order state-of-affairs constituted by instances of a law-making relation N between universals; then we get a modal claim like:

$$\text{Nec)}\Box(\text{N}(\text{F},\text{G})\rightarrow\forall\text{x}(\text{Fx}\rightarrow\text{Gx}))$$

Principles like (Nec) are famously ridden with problems. Supporters of DTA like Armstrong display some neo-Humean sympathies, and as such show some ambivalence with respect to the modal element in (Nec); the box operator patently does not appear in Armstrong's (1983, p. 97) formulation of the principle, in which it is only stated that laws "entail" or "involve" regularities. But the principle clearly has some modal value, if anything because otherwise the correlation between law and regularity would itself be a mere regularity (Bird, 2005, p. 150); although second-order instances of N are themselves contingent, it is not contingent that if N(F,G) then all Fs are Gs. But how is that not a non-Humean necessary relation between distinct existences, is unclear: the law is distinct from the regularity, after all.[5] Armstrong

[3]Hildebrand (2013, 2020), Tugby (2016), Bhogal (2020b, p. 164). Governance, mind you, is probably not the only way to draw the distinction between neo-Humeans and non-Humeans, albeit it is surely one which is rapidly gaining popularity in the literature. It seems to me that there is nothing keeping the two camps together but a loose family resemblance, or perhaps some vague background assumption about the nomic and/or modal status of reality. The basic idea, in the vaguest possible formulation, seems to be that non-Humeans believe there to be an irreducibly modal or nomic aspect to reality that cannot be factored into non-modal or non-nomic components. That said, it has been argued that certain neo-Humean accounts can preserve the "law-governed world-picture", as in Roberts (2008).

[4]The monicker "DTA" stands for the three historical supporters of the position: Dretske (1977), Tooley (1977), and Armstrong (1978, 1983).

[5]Armstrong never explicitly states as such, but it is a background assumption of his criticism of the regularity theory (Lewis, 1983, p. 366); furthermore, in Armstrong (1983, p. 86) he argues how difficult it would be to solve the problem by arguing that the law is simply the regularity plus something else; in that case the entailment and relative modality would turn into something less troubling: a conjunction necessitating a conjunct.

(1983, p. 86) characterizes this difficulty as "the puzzle of a relation between the universals which logically necessitates something distinct from itself: the uniformity". The problem, mind you, is not simply that such a necessary connection violates some neo-Humean dogma (although that certainly was an issue for a philosopher like Armstrong); the problem is that the source of this modality needs to be investigated, and unless the necessity is merely assumed as primitive, some explanation needs to be provided as to why N(F,G) is necessarily accompanied by $\forall x(Fx \rightarrow Gx)$; in other words: "I cannot see how it could be absolutely impossible to have N(F,G) and Fa without Ga" (Lewis, 1983, p. 366).[6] There are various ways to answer this question. Firstly, we may bite the bullet an simply accept brute necessary relations between distinct existences, as in Barker & Smart (2012, p. 716–717); secondly, we may want to claim that, broadly speaking, that there is something in the nature of N itself, as a necessitating universal, that warrants a modal principle such as (Nec); of course saying it doesn't make it so, in the sense that DTA supporters still have to show in what way N introduces such modality in the world; in fact, this was Lewis' concern about DTA. Neither solution comes for free: the former is thoroughly uninformative; whereas the latter contradicts Armstrong's position that all properties and relations, N included, are categorical and thus non-modal in character.[7]

At any rate, the general gist of DTA is that laws, as distinct from regularities, entail or necessitate them, and thus mentioning certain laws of nature constitutes an excellent explanation as to why reality is uniform in such-and-such a way. Laws of nature so understood as occasionally called "governing laws" starkly opposed to the merely descriptivist conception of law-statements of the regularity views proper of neo-Humeans, according to which laws themselves are nothing but a privileged subsets of regularities constituting the primitive distribution of properties across space-time. This divide between governing laws and laws-as-regularities is one of the deepest divides in contemporary metaphysics of science. In fact, it is paired with laws of nature that the very notion of governance takes shape, perhaps starting – historically speaking at least– with the theologically-driven conception of a nomically ordered world as a kingdom regulated by a king,[8] an impure analogy if anything because the modality of the laws of nature is alethic, whereas the modality of human laws is normative. Armstrong (1983, p. 106) was perhaps one of the first critics of neo-Humean laws to explicitly recognize that the analogy, as impure as it is, does have some bite; unfortunately, it is far from clear what exactly how

[6] Van Fraassen (1989, p. 64) called it the "inference problem", the inference being: "It is a law that A; therefore, A".

[7] A far more complicated route undertook by Armstrong (1983, p. 88–89)consists in claiming that N (F,G) is both a second-order state-of-affairs and itself a universal instantiated by pairs of first-order state-of-affairs such as F(a) and G(a); although how exactly that would solve the problem is unclear. For a criticism of the treatment of governance as instantiation, see Mumford (2004, p. 148–149). Finally, a primitivist route could be selected by the supporter of DTA, as proposed in Schaffer (2016a). I will investigate this proposal, with respect to dispositions, in the next chapter.

[8] Loewer (1996), Beebee (2000, p. 580–581), and Mumford (2004, p. 201–205).

(Armstrong's very resistance to the addition of the necessity operator in (Nec), as above, is telling in this regard). A necessary relation is the first way to cash out governance, but with the recent hyperintensional revolution came the idea that the perhaps a more-than-modal notion was needed to get the job done. In general, I agree with the sentiment: if there is a necessary relation between two items, we can still wonder about the nature of the relation that is the source of this modality. Analytic philosophers have a venerable example to draw from: identity as a necessary relation. That necessarily Clark Kent is co-located with Superman requires an explanation, which can be provided by claiming that Clark Kent is identical with Superman: a necessary relation cannot be the end of the story. In this case, identity is: but in other cases, other relations may be a source of modality in the world (e.g., parthood could also explain necessary co-location).

In the case of laws of nature, or any other governing item for that matter, the idea is that their presence is what makes, or produces certain regularity patterns instead of others, and it is what ultimately makes the world the way it is, instead of another; the relevant notion of dependence, or production, at hand here is intended to be non-metaphorical, and presumably non-causal.[9] Incidentally, keep in mind that DTA is not the only kind of position with governing laws: laws play a governing roles in the primitivist conception of laws in Carroll (1994) and Maudlin (2007), as they do according to Lange (2009) – amongst others; therefore, it is not the only position in which problems such as the ones above arise, as detailed in Schaffer (2016a).

In fact, governance is not limited to laws of nature at all. As hinted before, there may be other items playing a governing role in a non-Humean metaphysical system. Amongst other things, dispositions can too (Barker, 2013; Hildebrand, 2020). Bird's (2007a, p. 46) very derivation of regularities from dispositional essences was a very simple way to achieve this result, putting forward dispositional essentialism as the idea that things behave in certain way in certain circumstances ultimately because of the nature of the potencies they instantiate.[10] Cognate positions provide similar explanations, as in Tugby (2016). This is where we circle back to Dispositional Reality. How is governance to be cashed out in our framework? Two crucial questions need to be answered here:

A) What exactly is being governed, and what is doing the governing? In other words, what are the items to be found at the two ends of an instance of governance?
B) What kind of explanation is governance? And what relation does it bear with other, perhaps more familiar kind of explanations usually involved in this context, such as scientific or causal explanation? Because our account does not

[9]For an exception to the latter, see Maudlin (2007); but we will come back later to the relation between governance and causation.

[10]I'm using the term "potency" in Bird's sense: to refer to essentially dispositional properties. Technically, Bird (2007a, p. 46) is self-described as a derivation of laws of nature, instead of regularities; but the derived item is simply a law-statement describing a non-accidental regularity.

want to presuppose the reality of properties, and thus the reality of relations, to continue along the lines drawn so far we will cash out governance through an operationalist perspective; but the available data is far from sufficient to establish the nature of governance, both formally and informally.

We must answer both questions if we want to get a better idea of the kind of explanations provided by our variant of dispositionalism. As we will see, they are strictly related, but we must start somewhere; so will start with the first question, and then turn our attention to the second.

7.3 Exploring Governance

7.3.1 The Relata of Governance

What are the relata of governance? Assuming an operationalist perspective, instances of governance will take the form of "…because$_G$ …" completions, where "because$_G$" is our variably polyadic (on the right-hand side) operator of governance. But what kind of sentence can we find in such completions? On the governing side, dispositions: given the stance expressed in the beginning of this chapter, and the possibility of Dispositional Pluralism considered in the previous chapter, the sentences on one side will be dispositional ascriptions of any logical form we may wish.

What about the governed side of governance claims? Albeit not always recognized, non-Humeans often oscillate between two interpretations of governance. On the one hand, as Armstrong (1983) makes exceedingly clear, DTA involves the claim that laws of nature entail, necessitate, and explain wholesale regularities, traditionally expressed through universally quantified conditionals such as $\forall x (Fx \rightarrow Gx)$; on the other hand, he also claimed that "laws of nature govern particular states of affairs" (Armstrong, 1983, p. 106). Both interpretations have been upheld; the former is certainly more popular, as in (*inter alia*) Hildebrand (2013, 2020), Tugby (2016), and perhaps even more explicitly Schaffer (2016a); yet the latter has not been entirely lost, as in Loewer (1996, p. 187) and Bhogal (2020a, p. 6). A logically hybrid stance is espoused in Barker (2013, p. 625), according to which to be governed are instances of DTA-style regularities, e.g. facts like $[Fa \rightarrow Ga]$ or $[Fb \rightarrow Gb]$.[11]

So, which one is it? It is important here to not be unduly reductive and try to paint all laws with the same brush. After all, some laws of nature appear to be governing the behaviour of individual entities (the usual dispositionalist's example, Coulomb's

[11] To explicitly recognize this ambiguity is Wilsch (2021), which has recently drawn a similar (and very useful) distinction between "direct" and "conditional" necessity exertion of laws of nature.

Law, falls in this category), while others supposedly govern the behaviour of large or very large systems, perhaps the universe itself (e.g., laws of thermodynamics). In such cases, it makes sense to think of the law as governing the regularity directly.

It is hard to provide a definite statement of which laws belong to which category, but harder still to deny that a broad distinction between such categories exists. On the face of it, such categories of laws, and corresponding dispositions, face different problems. The latter category faces a problem of individuation: it is difficult to individuate the disposition governing the behaviour of large system directly. It cannot simply be an ordinary disposition of a specific entity within the system, but something more elusive. This is not a new difficulty, of course; it was often discussed by friends of powers as the problem of individuating powers (and bearers thereof) grounding laws such as symmetry and conservation principles (Bird, 2007a). This is where power metaphysics sometimes becomes more speculative in nature, e.g. the "world-dispositions" tentatively suggested by Vetter (2015) and Azzano (2022), or the possibility of "ultra-grounding" (Harré, 1986) such that some disposition is "partially grounded in a property or property-complex of [the object]'s whole physical environment" (Bauer, 2011, p. 97).

The former category of laws is perhaps more ordinary, and has an easier time identifying the relevant dispositions (e.g., the dispositions of electric charge, in the case of Coulomb's Law). But it faces a very different problem: understanding how governance works, with the mediation of the law-instance between the governing disposition and the overall regularity. In these cases what is doing the governing – whatever that is – does not explain or generate the regularity directly, all in one go: it does so bit by bit, explaining all of its instances, one at a time. It is because, say, for each entity in the domain that is F, the governing item makes it so that it is also G, that such an item explains the fact that the Fs are Gs. This is closer to the notion of governance expressed in Barker (2013, p. 625):

> X holding of F and G determines that if F is instantiated by object x then G will be instantiated, (or will tend to be), by x or something related to x.

"X" is understood by Barker as a second-order relation between properties F and G, either as a law of nature *per* DTA, or as part of the essence of a potency, as in Bird (2007a, 2007b). The identikit of this governing item X is not yet terribly important now; the crucial point is rather that for Barker governance relates some governing item X with individual instances of regularities.

I would like to focus on this more ordinary category of laws and dispositions, which govern a regularity only indirectly, if at all: more pertinently, they govern instances of regularity. This may seem simple enough, but it generates a different problem: what is the instance of a law? There seems to be some degree of ambiguity concerning the notion of a law-instance, oscillating between two different interpretations of the term: on the one side we have a scientific/confirmation sense of instance, and on the other, what we may call a logical sense. The scientific interpretation of a law-instance is perhaps the most widespread; according to Emery (2019, p. 1537–1538) presenting a "fairly natural" reading of the expression:

> an instance of such a law is a sequence of events such that, given the first event in the sequence, the law determines, as a matter of nomological necessity, that the later events in the sequence will occur. (So the relevant events will be events of some closed system being in some state.)

E.g., in the case of a causal law, an instance in this sense will be an individual causal sequence. This has also to do with instantiation: an instance of a law is any positive-instantiation pattern that might function as a confirmation of the law itself. E.g., an instance of Coulomb's Law might be constituted by two particles with a certain charge and at a certain distance exerting a force perfectly matching what is predicted by the law given the circumstances.

Albeit not always recognized, there is also a logical notion of instance. According to the simple regimentation employed in much of the literature referenced so far, law-statements are expressed at the first-order as universally quantified conditionals, and so are regularities. Usually, this orthodoxy amounts to something of the form: $\forall x (Fx \rightarrow Gx)$. When understood according to the orthodoxy, a law-statement $\forall x (Fx \rightarrow Gx)$ in which the quantifier ranges over a domain $\{a_1 \ldots a_n\}$ states that each of the $a_1 \ldots a_n$ satisfies the open formula $Fx \rightarrow Gx$. By way of simple quantifier elimination, logical instances of the law would be things like $Fa_1 \rightarrow Ga_1$, $Fa_2 \rightarrow Ga_2$, and so on. Such logical instances express what the law is stating: viz., that each F is also a G, either trivially or non-trivially. But sentences such as these do not satisfy Emery's definition, and thus do not count as instances in the scientific sense. The reader might hereby be reminded of the Paradox of Confirmation; something not being F (trivially) satisfies $Fx \rightarrow Gx$, but does not provide a sequence of events as described by the law, and thus does not confirm it. A scientific instance in this sense will probably be something like $Fa_1 \& Ga_1$; again in Emery's words, a sequence such that the law "nomically necessitates" that, given the former, the latter also obtains (in our previous example: two particles with a certain charge and at a certain distance exerting a force matching what the law predicts). If $\forall x(Fx \rightarrow Gx)$, then each of the $a_1 \ldots a_n$ satisfies $Fx \rightarrow Gx$, but only those that satisfy it non-trivially also satisfy $Fa_1 \& Ga_1$. Hence, "scientific instances" and "logical instances" are materially different: only a (possibly empty) subset of logical instances of a law also are scientific instances. In other words, what we are expressing with the regularity $\forall x(Fx \rightarrow Gx)$ is not simply that *some* things are involved in a sequence from F to G; and surely not that *all* things are involved in a sequence from F to G. The former is incomplete, and the latter false. When expressing a regularity $\forall x(Fx \rightarrow Gx)$, we are saying that, of all things, either they are involved in a sequence from F to G, or they are not F at all, which is, truth-functionally, not very far from saying that they satisfy $Fx \rightarrow Gx$.

What should we make of this? To my mind, there are (at least) two readings of this situation. Reading (1): we could read this situation as proof that law-statements shouldn't be expressed as universally-quantified conditionals in the first place: for one, Armstrong (1983, p. 43), otherwise a supporter of the orthodox stance, tentatively suggests this route when discussing the Paradox of Confirmation; in his words, "'$\forall x(Fx\rightarrow Gx)$' cannot be substituted for 'it is a law that Fs are Gs'". Reading (1) has independent motivation; after all, that "$\forall x(Fx \rightarrow Gx)$" satisfactorily formalizes law-statements is a very controversial topic on its own. The broadest worry is whether such a simplistic approach can actually model scientific laws, as opposed to the simple regularities one might be introduced to in an introductory course on philosophy of science, such as "all ravens are black", or "all uranium spheres are less than a mile in diameter". Maudlin (2007, p. 11) famously aired this worry:

> The fundamental law of Newtonian mechanics, the mathematical consequence of Newton's first two laws, is $\mathbf{F} = m\mathbf{a}$ or $\mathbf{F} = m\, d^2\mathbf{x}/dt^2$ or, most precisely, $\mathbf{F} = md(mv)/dt$. The fundamental law of non-relativistic quantum mechanics, Schrödinger's equation, is $i\hbar\, \partial\Psi/\partial t = \hat{H}\Psi$. No doubt these can be tortured into a form similar to $\forall x(Fx \rightarrow Gx)$, but it is hard to see what the purpose of the exercise would be.

This is a cogent criticism, albeit one which can be resisted, e.g. Friend (2016). Furthermore, the fact remains that laws of nature are supposed to describe a universally applicable sequence of behaviours of certain entities or regions of space-time, as Maudlin *(ibid.)* himself admits that "[w]hat is most obvious about these laws is that they describe how the physical state of a system or particle evolves through time. The laws are generally presumed to be universally valid". Maudlin's following point, that such features are not directly built into the mathematical structure of the law itself, does not mean that laws couldn't expressed in this fashion, perhaps for the philosophical purpose of highlighting the above features (as it is the case here), but only that such formulation may end up being not scientifically or mathematically privileged.

Reading (2): we could take our distinction at face value, and claim that the expression "instance" unpredictably oscillates between a logical meaning and a scientific/confirmation-based meaning.

It is hard to say which option is the right one, between (1) and (2). But methodologically speaking, I think it would be best to select (2). While the jury might still be out whether the orthodoxy on the logical form of law-statement is right, one can never go wrong drawing a finer-grained distinction: if the distinction turns out to be redundant, no harm done; but if it isn't, this will be a crucial advantage of our position. Therefore, let us distinguish between "logical instances" of regularities, which I will call L-instances, and "scientific instances" of regularities, which I will call S-instances. On the orthodox formalization of law-statements and regularity-statements, an L-instance is something along the lines of $Fa_1 \rightarrow Ga_1$; an S-instance is something like $Fa_1\&Ga_1$. Only some L-instances are also cases of S-instances, for the simple fact that only in some cases if $Fa_1 \rightarrow Ga_1$ is true then $Fa_1\&Ga_1$ is also true.

Finally, we can turn to the question: what kind of instance is governed by dispositional ascriptions? L-instances or S-instances? It seems to me that, given some premises, the right answer is L-instances. The crucial premise is that governance, whatever kind of explanation it is (more on that below), entails some kind of necessitation, similar to the kind of (Nec) principle espoused by supporters of DTA. If such a necessitation is in place, however, it becomes clear that only L-instances are necessitated by disposition ascriptions. S-instances are not. Remember that on the orthodox formalization of regularity statements $\forall x(Fx \rightarrow Gx)$, an L-instance is $Fa_1 \rightarrow Ga_1$, while an S-instance is $Fa_1\&Ga_1$. That some x is disposed to G if F may govern, and thus necessitate, that $Fx \rightarrow Gx$; but it does not necessitate that

Fx&Gx. This is for the simple reason that not all things disposed to behave in such-a-such a way are actually behaving in such and such a way: dispositions may be dormant. This is not a unique feature of my account. After all, Emery's quote above states that, given a sequence that constitutes a S-instance, the law determines/necessitates that *given* the first part of the sequence, the latter *also* occurs. Barker (2013, p. 625) also claims that the disposition determines that *if* F is instantiated *then* G will also be instantiated. All of these approximate something in the ballpark of Fx → Gx, thus, an L-instance. Things are similar in e.g. Bird's (2007a) account of potencies whose essences are governing regularities and their instances (both conveniently formalized according to the orthodox formulation): that solubility is a potency with a dispositional essence may necessitate that a soluble thing, if immersed in water, dissolves; it doesn't necessitate that every soluble thing is actually immersed in water and dissolving. Bird's potencies, just like regular dispositions, may very well be dormant.[12]

[12]There is independent motivation for the claim that dispositions are never dormant, and are in fact constantly manifesting. This move has the potential to collapse the distinction between L-instances and S-instances, and ultimately, the orthodox regimentation of law-statements *qua* conditionals.

Consider for example electric charge; we may want to characterize the dispositions associated with electric charge as a disposition to attract or repel other charged particles when a certain distance – the "when" specifying the stimulus of the disposition–, but this is quite an awkward way of characterizing the disposition. A particle, by virtue of being charged, is always doing something, namely, inducing an electric field. The disposition is never truly dormant. Thus, Coulomb's Law shouldn't be understood as a conditional dictating what charged particles would do if placed in a certain scenario, but rather as describing a constantly evolving process governed by the dispositions of a charged particle; on this picture, the difference between L-instances and S-instances of Coulomb's Law is not so clearly demarcated, and, depending on how the law is effectively expressed, may very well disappear altogether. Another case, perhaps a little more controversial, is radioactive decay: some materials are known to emit ionizing radiations because of subatomic instability, and for that reason are said to be radioactive. Radioactivity does not need to be activated, and there is no sense to be made in thinking about it as a dormant disposition: as long as it is possessed (as long as the atoms are unstable), there is a spontaneous and continuous emission. Examples such as these can be generalized to the conclusion that the dispositions of fundamental physics are in fact spontaneously and constantly manifesting, as in Esfeld and Sachse (2011, p. 37–38).

Even if that were true, however, this point hardly generalizes. Beyond the dispositions of fundamental physics there are explanatorily relevant dispositions which indeed seem to require stimuli, and for which the distinction between L-instances and S-instances is very much relevant. Some of these appear to be explanatorily relevant dispositions of special sciences, and, as such, excellent candidates for Dispositional Reality. E.g., developmental or phenotypic plasticity and robustness are crucial dispositions in current evolutionary developmental biology, usually understood as provided with a stimulus as well as a manifestation; according to the account in Austin & Nuño de la Rosa (2018, p. 476), "alterations in the environmental conditions and thus heterochronical and heterotopical alterations in the upstream signaling of these systems (that is, changes in their *stimulus* conditions) can result in downstream qualitative alterations – in shape, size, pigmentation, etc. – of its phenotypic product (that is, its *manifestation*)." In conclusion, I will stick to the safer options that dispositions can go dormant, and thus govern L-instances formulated with the help of conditionals.

7.3.2 *The Nature of Governance*

We have shed some light on the relata of governance. But what about its nature as an explanatory relation? In what sense of "explain" does a thing being disposed explain its behaviour?

The kind of clarification I have in mind is not meant to provide a formal theory concerning governance: any such formal theory always springs from a certain informal grasp on the nature of the relevant item, e.g. its behaviour with respect to logical operators (as in the case of the logic of ground provided by Fine). This kind of informal grasp is exactly what we are trying to achieve for governance here, vis-à-vis other types of explanations.

This is a tough issue to tackle, if anything because in the zoo of explanation-types we can draw from, governance doesn't resemble any of them: it is far easier to say what governance is not, than what governance actually is. But there's value in that too. It seems to me that governance is not scientific explanation. Governance is not constitutive explanation either. And finally, governance is not causal explanation. Let's see these points in turn. Then we can try to positively say what governance is.

7.3.2.1 Governance Is Not Scientific Explanation

Well-read readers in philosophy of science may have reacted in the following way to the previous sections: well, of course laws of nature explain certain phenomena, such as individual causal patterns or more general regularities. This is what scientific explanation is all about: is this talk of "governance" a fancy way to talk about scientific explanation? Unfortunately, there is no place within this book to present an exhaustive theory of scientific explanation. But a generic grasp on the notion is, in my opinion, enough to answer the previous question in the negative: governance is not the same as scientific explanation.

To be clear, it surely exists a metaphysically non-committal reading of governing/obeying locutions, to the extent that a law "governs" a phenomenon simply in the sense that it applies to it; conversely, a certain phenomenon "obeys" the law simply in the sense that the evolution of that phenomenon fits what the laws says. Scientists are known to employ this reading too; e.g. Mumford (2004) points out an oscillation on Richard Feynman's part between a descriptivist and heavyweight conception of laws of nature that does not prevent him to claim that laws "govern" phenomena, or contrariwise that phenomena "obey" laws – perhaps in a purely metaphorical sense.

Under this non-committing reading of the expression, governance may very well equate with scientific explanation. After all, that of scientific explanation is a malleable concept that can be accommodated even in the most descriptivist neo-Humean frameworks, perhaps as a merely epistemic or systematizing phenomenon related to deduction.[13] In fact, that laws of nature *scientifically* explain phenomena is the crux of a popular neo-Humean response to the explanatory circularity objection in Loewer (2012). But such descriptive laws do not govern such phenomena in the more robust sense usually meant by non-Humean metaphysicians.[14] For them, it is not simply the case that laws of nature (or some other item) "govern" regularities in the sense that they apply to them, e.g., by saying that law-statements describe the behaviour of certain things; they make the stronger claim that things behave in the way they do exactly *because* certain laws apply to them, in a sense of "because" we are trying to define. To use more risqué terminology, which may or may not be vindicated by a mature analysis of governance: things behave according to the law because something is *compelling* them to. This is what a non-Humean pushing-and-pulling sounds like, and it is a reading of "governing" more metaphysically committing than the one introduced before. On this reading of the term, most neo-Humeans believe that laws scientifically explain regularities, but do not govern them.[15] Thus, while I suspect that there is an interesting, and possibly very intimate relation between governance and scientific explanation, a study of governance cannot be exhausted by the study of scientific explanation.

7.3.2.2 Governance Is Not Constitutive Explanation

Perhaps governance is a kind of constitutive explanation, then? In the previous chapter I introduced the notion of constitutive explanation, viz. as a variably polyadic operator which entails instances of (also variably polyadic) generalized identity. Given arbitrary sentences p and q, the constitutive character of this kind of explanation approximates to something like.

Ce) p because_c q $\rightarrow$ p $\equiv$ q

Based on this principle, it is safe to say that governance is *not* constitutive explanation. Not only because, as above, we already have a "constitutive base" of such regularities, and thus the two explanatory bases would somehow come into

---

[13] Loewer (1996, p. 113), Beebee (2000, p. 578), Miller (2015, p. 1326), and Bhogal (2020a, p. 6). For a contrast, see Dretske (1977, p. 262).

[14] Famously, one may harbour worries about the explanatory function of descriptive laws in a neo-Humean system, even in this deflated sense of explanation. As descriptive laws naturally emerge as a theoretically virtuous subset of regularities across the space-time manifold, the only kind of explanation such laws can provide is *post-hoc*: the genuine laws of nature will not manifest themselves until time is over, so to speak: thus, when it comes to the individuation of laws as explanatory, descriptivists must perform a methodologically impossible "wait until the end of times" manoeuvre. Most notably, prediction (which on the DN model is but another facet of explanation) becomes impossible: for at the end of times, there's nothing left to predict. In other words, neo-Humeans must accept the problem of induction as ultimately unsolvable from a metaphysical point of view: every allegedly reliable prediction is entirely conditional on a past pattern extending into the future, something for which there is no assurance.

[15] For a famous exception, again see Roberts (2008).

conflict; but, more clearly because of clause (Ce) characterizing constitutive explanation itself. It seems awfully wrong to claim that governing dispositions, or governing laws, *just are* regularities. They are patently not: in fact, the very reason why the DTA theory of laws was proposed, e.g. as in Armstrong (1983) was to oppose regularist theories of laws in which laws ultimately amounted to regularities: laws of nature as higher-order relations between universals or states of affairs are *not* regularities. Something similar goes for Bird's (2007a) dispositional essentialism, according to which, with some simplification, soluble things dissolve in water by virtue of the nature of solubility; but obviously solubility having a certain nature is not identical to soluble things dissolving in water – this last claim being so wrong that it approximates a category mistake: it is one thing for a certain potency to have a certain nature, and another one for things to behave according to that nature. The same goes for our account of governing dispositions. After all, the very reason why non-Humean accounts escape Maudlin's (2007, p. 72) circularity problem, successful or unsuccessful as it might be, is because on such non-Humean accounts items such as governing laws are ultimately not constituted by the regularities they govern. Relatedly, if grounding entails reduction, governance is not grounding either.[16]

In conclusion, given the informal characterization of the notion employed so far in the literature, it is safe to say that governance is not constitutive explanation: because$_G$ is not because$_C$.

7.3.2.3 Governance Is Not Causal Explanation

The intuitive grasp on the notion of governance presented so far involves the idea that, when an instance of a causal regularity is produced, it is somehow explained, or produced, or "compelled" – to reiterate the suggestive but metaphysically vague expression used before– by some governing item, be it a heavyweight law, the essence of a potency or what have you. I suspect that, pre-theoretically, this is a widespread intuition as well, which explains why a targeted intellectual effort is required to move from the non-Humean camp to the neo-Human camp. Scientists as well as philosophers clearly share this intuition, which is why the non-Humean option may strike one as *prima facie* appealing. Davies (2007, p. 20) talking about the specific case of laws of nature, commented that:

> [t]he idea of laws began as a way of formalizing patterns in nature that connect physical events. Physicists became so familiar with the laws that somewhere along the way the laws themselves – as opposed to the events they describe – became promoted to reality. The laws took on a life of their own. It is hard for non-scientists to grasp the significance of this step. [T]the laws of physics are said to inhabit an abstract realm and touch the physical world only when they "act." It's almost as if the laws are lying in wait, ready to seize control of a physical process and compel it to comply [. . .]. This "prescriptive" view of physical laws as having power over nature is not without its detractors (namely, philosophers who prefer a

[16] According to Rosen (2010), grounding entails reduction; this is a somewhat popular stance, as claimed in Trogdon (2013), but not without criticisms, as in Audi (2012).

> "descriptive" view). But most physicists working on fundamental topics inhabit the prescriptive camp, even if they won't own up to it explicitly. So we have this image of really existing laws of physics ensconced in a transcendent aerie, lording it over lowly matter.

This is exactly the kind of statement that a philosopher would find disquieting (one can perhaps detect a hint of irony in Davies' choice of words as well). David Armstrong, one of the first philosophers to explicitly talk about the governance of laws of nature, in Armstrong (1983, p. 86) discussed a very similar position as follows:

> [t]his is the picture which I get from the Phaedo. The Forms up in Heaven are related in a certain way, and, as a result, a uniformity is produced on earth. Such a doctrine is extremely mysterious. The mystery is only a little reduced by bringing the Forms to earth and letting them exist only in their instantiations. We still have the puzzle of a relation between the universals which logically necessitates something distinct from itself: the uniformity.

Although Armstrong's worries are clear (and clear is their source: neo-Humean philosophical sympathies), exactly which positive account is supposed to replace this "mysterious doctrine" is less clear. According to him, the DTA theory is supposed to restore the truth behind talk of governance, surely abandoning the metaphor as a philosophical guide, yet recognizing that the metaphor has some bite (Armstrong, 1983, p. 106). But how exactly that will pan out, is still unclear.

The crudest idea is to understand this "acting" and "compelling" of governing items as inherently causal in character; thus governance, as an explanatory notion, is to be subsumed as a special case of objective causal explanation, in which to be explained are individual causal sequences, or portions, or generalizations thereof. There is some philosophical resistance to this idea; according to Emery (2019), governing laws may explain their instances, but they do not cause them, a widespread position allegedly, as "the vast majority of philosophers are committed to the view that laws do not enter into causal relations" (Emery, 2019, p. 1545). The same could be said for other governing views, such as the one in Bird (2007a); we may derive regularities, or instances thereof, from the essence of potencies – but we shouldn't expect the latter to *cause* the former. Emery *(ibid.)* brings three points in favour of this conclusion, without much elaboration:

> The thesis that laws are relations between universals. (A relation between universals cannot cause anything.)
> The thesis that laws are regularities. (A regularity cannot cause an instance of a regularity.)
> The thesis that causal processes transfer some quantity from cause to effect. (A law does not transit a quantity to any of its instances.)

The first two theses are not in my opinion very conclusive, as they are too much - theory-dependent to be philosophically reliable to establish my point: whether governance is causation shouldn't require the claim that laws are relations between universals, or regularities (in fact – see below – I believe they are neither). But there is something to be said in favour of the third point, that requires further elaboration.

This is where regimenting laws and regularities according to the orthodoxy may lead us astray. Strings like "$\forall x(Fx \rightarrow Gx)$" are apt to formalize natural language sentences like, again, "all ravens are black"; but law-statements are different in one important respect: they have a built-in mathematical structure, in the sense that they relate quantities, expressed through numerical variables; causation, more generally, should be understood as connecting quantities. The same, incidentally, must be true of dispositions: if dispositions must play the role previously played by laws in a non-Humean metaphysics, they must display some sort of built-in mathematical structure, presumably in the stimulus and/or manifestation specification.[17] This is one of the reasons, as expressed in the previous chapter, to prefer determinable dispositional ascriptions to determinate ones.

What does that mean for our current purpose? Under the simplified assumption that "F" and "G" in an orthodoxically-expressed law-statement hide a more complicated mathematical structure, the only quantities affected in a thusly governed causal sequence are exactly those expressed in "F" and "G", and no more. To claim that there is further causation between the law itself that Fs are Gs, and F&G causal sequence would entail that there are further quantities affected by the process: yet this seems like pure metaphysical speculation, as there doesn't appear to be no non-ad hoc quantity to fit the requirements. The law itself (or the disposition itself) is not a quantity transferred or otherwise affected by its governing – at least, not in any observably measurable capacity: in other words, there is no observable causal action that corresponds to governing. The same goes for Bird's second-order essences of potencies: they are not quantities *per se*.

Again, to claim that such governing items directly take part in the causings they allegedly govern seems like a category mistake. It would appear more appropriate to consider such governing items as providing rules expressing how such causings should occur in the corresponding lower-order processes. They are more like a building manual, than an actual builder. This gives us some additional hint as to how governance should be formulated.

7.3.2.4 So, What Is Governance?

We have tentatively decided to express governance as a variably polyadic sentential operator "…because$_G$…" connecting, on the one side, dispositional ascriptions of some kind, and on the other, instances of law-statement or regularity-statement, understood as sentential expressions of causal sequences or other kinds of nomically-regulated evolutions of a system through time. As above, governance in a non-Humean metaphysical system is not scientific explanation, nor constitutive explanation, nor causal explanation. So, what is it?

To philosophically express the intuitive grasp that we have on the notion we could say that governance is a non-causal and non-constitutive objective explanation, characterized through an equivalent of the "generalized identity" for non-causal determination. In a simplified two-sentences scenario:

[17] As in Schrenk (2007) and Sartenaer, Guay & Humphreys (2021, p. 7–10).

Gov) If p $because_G$ q, then that q non-causally depends on that p.

The viability of principles such as (Gov) of course depend on what we take this operator of non-causal determination to be. Again, I don't think it will be excessively useful to focus on the formal properties, on the logic of this kind of non-causal dependence, but on a tentative grasp of the notion from which choices in this matter can be attempted. This non-causal dependence is to be understood as non-causal and non-constitutive dependence, presumably of the more-than-modal kind. But it still entails a modal relation; for this reason, governing items are sometimes understood as generating a restriction in the possible distribution of properties across the so-called mosaic, as in Hildebrand (2013).

For our purpose, we can take governance as being a *sui generis* kind of explanation; in other words: the fact that p explains, and thus determines that the fact that q takes place.[18] We are not completely in the dark, however: there are certain things that we can positively say about governance, which may help bring it into sharper focus. First of all, just like grounding and causation, governance is Janus-faced to the extent that it has both a determinative and explanatory dimension, codified by the (Gov) principle above; and similarly to grounding, we may legitimately wonder whether we should understand the relationship between this determinative and explanatory aspect along unionist or separatist lines, viz. whether governance, as a non-causal determinative relation also is itself inherently explanatory in character, or whether successful governing explanations merely track relations of non-causal determination.[19] Some details concerning this distinction are offered in Chap. 3; the reader will perhaps remember that the choice between unionism and separatism presupposes the doctrine of explanatory realism, viz. the idea that "C is an *explanans* for E in virtue of the fact that c bears to e some determinate objective relation R" (Kim, 2010, p. 149). To ask, for an arbitrary explanatory relation $because_X$, whether $\mathbb{S}(because_X)$ amounts to asking whether successful $because_X$ explanations map unto some portion of metaphysical structure; we can then wonder whether this is to be understood along unionist or separatist lines. An explanatory relation that we usually understand along realist lines is causal explanation, the "determinate objective relation R" in this case being causation itself. The kind of non-causal explanation more extensively discussed in this book are constitutive explanation. Constitutive explanation is not real, and the reason is the one presented in Chap. 3: because the reality of constitutive explanation would generate a problematic layering of reality; to the extent that metaphysical grounding entails reduction, this is my opinion on the reality of grounding as well. If, say, water being hot is constitutively explained by the water mean kinetic energy being high, this also mean that for water to be hot just is for the water to have high mean kinetic energy. These are not, so to speak, two

[18] Remember, once again, that fact-talk is dispensable. The above formulations, in (Gov) are grammatically awkward, but this is to be expected. There is some degree of artificial syntactic manipulation in Sider's (2011) "…is like when…" operator, and Rayo's (2013) "…just is…" operator.

[19] See Chap. 3 for more details on this distinction.

separate portions of reality: what rather happens is that we have two distinct ways to express the same fact; one more metaphysically perspicuous, and the other less. To hold the contrary opinion that that kind of explanation is real, means that reality is structured to separate "real facts" from the "less real" ones, shadowy facts whose reality is somehow reduced or borrowed from the ones they lay on top of. To circle back to Chap. 3, this is not my preferred metaphysical set-up.[20]

With this understanding in mind, we can ask: is governance real, viz. a part of metaphysical structure? In other words, $\mathbb{S}(\text{because}_G)$? The reader should remember that tinkering with the logical form of governance – shifting from a predicative to an operationalist formalization – does not by itself solve the metaphysical issue of the reality of governance, since the $\mathbb{S}$ joint-carving operator can attach to expression of any syntactic category. Naturally, this is where the methodology developed in this book comes into play again: if some theory provided with governance principles score more points vis-à-vis its neo-Humean adversaries, this will be good indication that there is indeed such a thing as governance, viz. that governance is part of metaphysical structure. This is of course a point that cannot be exhaustively settled in these pages. Yet, given the points made above, there is a point to be discussed concerning the reality of governance. If, assuming, governance is part of metaphysical structure, doesn't that mean that its introduction will generate an unacceptable layering of reality, such as the one I criticized in Chaps. 2 and 3? The answer is: no. To motivate this answer, one must remember that we are not dealing with constitutive explanation here; in other words, if p because_G q, then the fact that p is something-and-above the fact that q. In this respect, governance is more like causation: the existence of causal relations does not generate layers of "more-real" or "less real" causal relata: that's because causal explanation too is non-constitutive: what is caused and what is doing the causing are distinct, equally real occurrences. The same goes for governance.

That said, I still believe that the indirect methodology for metaphysics set up in Chap. 3 I still the best way to approach the epistemology of governance, as I suspect that there is no privileged epistemic access to non-causal dependence relations as in (Gov). The resulting picture can be summarized as such: there are pushings and pullings in the world, all of them non-Humean to the extent that they connect distinct portions of reality (which is why, as hinted before in this book, grounding as constitutive explanation is not inherently non-Humean); some of these pushings and pullings are causal in character, and some of them are not; governance is amongst the latter.

This point leads us to a final observation that we can make concerning governance and its relation with causation. If I am on the right track, governance is not causation, although it shares certain features with it; thus one may wonder whether we should understand this similar relations under a more unified framework, perhaps by claiming that causation and governance are two species of a single genus.

[20] In this matter, I agree with Rayo (2013) that the kind of "just is" generalized identity statements entailed by constitutive explanations, mostly exist to close theoretical or conceptual gaps.

A similar claim has been put forward concerning causation and grounding, so perhaps we could attempt a similar move here and claim that governance is metaphysical causation.[21] Such a bold move is far beyond what I am expecting to achieve in these pages; and furthermore, I am not even sure that such a unification is a desirable goal to achieve.

There is something more to be said however, about the intimate relation between causation and governance. As we have seen there are reasons to believe that a governing item is not an additional cause to obtain a certain effect, but rather something that regulates how a certain causal process evolves through time; in the simplest sense, this is enshrined in the mathematical structure displayed by the governing item, which regulates variations of the relevant quantities. Roughly put: a governing item, be it a law or a disposition, should not be understood as a further quantity affected in the causal process, or a further cause pushing the process along its way – and more of a blueprint as to how the process should evolve through time. This is what is effectively missing in a neo-Humean metaphysics. This vindicates our previous approach in which "...because$_G$..." sentences display, on the left, individual L-instances of law-statements or regularity-statements, which describe the evolution through time of individual causal processes. In a slogan: governance is not causation, but governance governs causation.

7.4 Explaining Regularities with Dispositions

It is finally time to formulate an account of governance within our framework of Dispositional Reality. In this sect. I will present the account, and then discuss a problem emerging from it.

7.4.1 *The Account*

According to the metaphysical picture advocated here, there are determinative *cum* explanatory relations in the world; some of these pushings and pullings are causal in character, and some of them are not; governance is amongst the latter. Governance is a determinative relation with an explanatory character which is both non-causal and non-constitutive; whether it is to be understood along unionist or separatist line, I will not say. Being non-constitutive, it does not layer reality in a problematic way. Furthermore, it has an interesting correlation with causation and causal explanation,

[21] As in Schaffer (2016b) and Wilson (2018); a slightly similar position, as in Bennett (2017), is that grounding and causation are two species of a "building" genus. See Bernstein (2016) for criticism. To the extent that grounding-based explanations are constitutive, whereas governing-based and causation-based ones are not, it seems to me that governance, more than grounding, ought to be affiliated with causation.

in the following sense: one or more dispositional ascriptions govern individual regularity-instances, thus somehow regulating that causal sequence or process (including the mathematical structure of quantitative changes).[22] How exactly that happens, it is difficult to say: I leave governance as a *sui generis* relation for now (but I will have something more to say about this in next chapter). I have also not specific *what* exactly the dispositions in the *explanans* are predicated to, be it some entity or region of space-time interested by the causal process in the *explanandum*; in this matter, flexibility is clearly preferable. Although governance is not an inherently modal relation, it presumably entails a kind of necessitation as well, presumably a dispositional equivalent of the (Nec) principle for DTA. As governance entails necessitation, and furthermore is not constitutive explanation, the view presented so far squarely falls within the non-Humean camp: for it clearly envisions necessary relations between distinct existences.

Once we tentatively assume the orthodox formalization of law-statements and regularity-statements as universally quantified conditionals, the simplest dispositional predications we can use to effectively formulate instances of governance along the lines proposed above are, in fact, of the simplest stimulus-manifestation kind. For simplicity let us use the "D…-…" predicate-modifier that, for any two arbitrary one-place predicates P_1 and P_2 provides the further one-place predicate $D_{P_1 \square P_2}$, which in the intended interpretation is to be read as "disposed to P_2 if P_1", where P_1 is the stimulus specification predicate, and P_2 is the manifestation specification predicate. Thus, given a non-accidental regularity such that $\forall x(Fx \rightarrow Gx)$, such that the quantifier ranges over a domain $\{a_1 \ldots a_n\}$, an instance of governance will be something along the lines of

1) $Fa_1 \rightarrow Ga_1$ because$_G$ $D_{F\text{-}G}a_1$

This entails, because of (Gov).

2) $Fa_1 \rightarrow Ga_1$ non-causally depends on $D_{F\text{-}G}a_1$

And also, by way of necessitation:

3) $\square(D_{F-G}a_1 \rightarrow (Fa_1 \rightarrow Ga_1))$.

This account allows us to recapture, amongst other things, the modal robustness of nomic regularities vis-à-vis accidental ones: in fact, a non-accidental regularity is here understood as one whose L-instances are governed by dispositions: thus, as long as those dispositions are present, the regularities also occur.

Two observations may naturally occur. Observation number one: our previous description of the stimulus conditions by way of a simple letter "F" or "G" may hide an important complication. This is another way in which orthodoxy may lead us

[22] To be clear, I will not assume that all laws of nature are causal in character (the standard example here are symmetry or conservation principles, which will require a separate treatment), and thus I do not assume that all non-accidental regularities are causal in character. For now let us concentrate on these cases, viz. causal regularities whose instances are individual causal sequences or processes.

astray. On the contrary, stimulus conditions are known to be a far more complicated affair; amongst other things they will probably mention certain properties, and, more specifically, the kind of properties that, as we have discussed in the previous chapter, are necessarily related (albeit distinct) from the original disposition whose stimulus we are currently observing. Consider one of the dispositionalists' favourite toy-example, viz. that of solubility. Assume that it is a causal regularity that soluble things dissolve when immersed in water; when formulated according to the orthodoxy, the antecedent will presumably mention solubility as well. It is not all things dissolve when immersed in water, but only soluble ones. And so, the disposition distinct yet necessarily related to solubility (as explained in the previous chapter), must mention solubility in its stimulus condition, if governance is to be formulated according to the principle (1) above. And the same in many other cases: if we really wish to torture Coulomb's Law in conditional form, charge quantities must be mentioned in the antecedent and thus in the stimulus of the disposition to attract and repel according to the law, a disposition whose possession by the particle is necessitated by the possession of charge itself. This may seem like an idle observation, but it will prove to be important shortly.

Observation number two: what about the entire regularity? We have so focused on the claim that disposition instances govern L-instances of regularities, but what about the regularity these are L-instances of? After all, the intuitive grip of a non-Humean metaphysics which envisions governance is the idea that the global distribution of properties across space-time, in other words, the "mosaic", is not brute or unexplainable, but is determined by something: after all, "non-Humeans think that regularities are imposed on nature" (Hildebrand, 2020, p. 2).

There's an interesting story to be told here as well. As a matter of fact, although dispositions do not directly govern such regularities on my account, there is something to be said about the modal relation between them. How? We can see that there is a ready-made explanatory story concerning regularities: viz. the constitutive explanatory relation going from L-instances to the universally quantified sentences, which can be expressed through a standard logic of grounding as in Fine (2012). Why is it that $\forall x(Fx \rightarrow Gx)$? Because$_c$ $Fa_1 \rightarrow Ga_1 \ldots Fa_n \rightarrow Ga_n$ plus a totality fact that the $a_1 \ldots a_n$ are all there is. Thus, L-instances constitutively explain (collectively, plus a totality fact) the regularity. Dispositions have nothing to do with this explanation, *per se*. But, as we have seen, dispositions govern L-instances of the regularity. Now, because$_C$ and because$_G$ are different kind of explanations, we cannot exactly chain them together to claim that dispositions directly explain regularities. But we can say something weaker, while still non-trivial. Both explanations entail a modal relation of necessitation,[23] and necessitation is transitive: thus, we can safely reach the conclusion that dispositions necessitate regularities. This is a non-trivially non-Humean claim to make; on a neo-Humean metaphysical systems, how things

[23] That constitutive explanation entails necessitation is a claim that draws strength from the standard assumption that grounding entails necessitation. There are dissenting opinions, however, as in Leuenberger (2014) and Skiles (2015).

are disposed and how they are propertied (and thus what the global distribution of properties will be) are independent variables. Not so much on the present account.

The complete account presents a couple of additional complications, and is thus worth spelling out. Let us call "S" our stimulus specification and "M" are manifestation specification; and let us call P the property usually associated to the stimulus and manifestation (as above). The resulting regularity will take the traditional logical form of $\forall x(Sx \ \& \ Px \rightarrow Mx)$; e.g., an electrically charged particle in a certain environment will exert such-and-such a force. The relevant disposition will, very simply, be something along the lines of $D_{P\text{-}M}$. So, how do we get the regularity from the properties and the dispositions? Let us work with a toy-world where $\forall x$ $(S\&Px \rightarrow Mx)$ is true, and is furthermore, a non-accidental regularity, and where the quantifier ranges over the domain $\{a_1, a_2, a_3\}$. There are several cases here: things that are both S and P, and thus M (Case 1); and things that are not both S and P (thus making the relevant L-instance of the regularity trivially true); of those, certain things are not S (Case 2), and others are not P (Case 3). For illustrative purposes, imagine that Case 1 is represented by a_1, Case 2 by a_2, and Case 3 by a_3. Now, let us reconstruct the explanatory and necessitation relations in this toy-world.

Case 1 and Case 2 are fairly similar, as both a_1 and a_2 are both P. Furthermore the property P that (by virtue of the previous chapter) non-constitutively explains and necessitates the disposition $D_{S\text{-}M}$. So we have:

4) $\Box(Pa_1 \rightarrow D_{S-M}a_1)$
5) $\Box(Pa_2 \rightarrow D_{S-M}a_2)$

By the Axiom T of modal logic

6) $Pa_1 \rightarrow D_{S-M}a_1$
7) $Pa_2 \rightarrow D_{S-M}a_2$

Furthermore, we know that by claim (3) above, the dispositions govern and necessitate L-instances of regularities. Thus, again by Axiom T:

8) $D_{S\text{-}M}a_1 \rightarrow (Sa_1 \rightarrow Ma_1)$
9) $D_{S\text{-}M}a_2 \rightarrow (Sa_2 \rightarrow Ma_2)$

Furthermore, by the transitivity of the material conditional, from (6) and (7) we also get:

10) $Pa_1 \rightarrow (Sa_1 \rightarrow Ma_1)$
11) $Pa_2 \rightarrow (Sa_2 \rightarrow Ma_2)$

And by simple propositional logic:

12) $Pa_1 \ \& \ Sa_1 \rightarrow Ma_1$
13) $Pa_2 \ \& \ Sa_2 \rightarrow Ma_2$

These are the first two complete L-instances of the overall regularity. Case 3 is different, as most of the above explanations do not take place in that case: object a_3 is not P; a_3 does not have the property, nor the disposition, and thus there is no governing going on. However, exactly for that reason, $Pa_3 \ \& \ Sa_3 \rightarrow Ma_3$ is vacuously true. There's only one step left now, and it is the following. Pa_1 & $Sa_1 \rightarrow Ma_1$ *plus* Pa_2 & $Sa_2 \rightarrow Ma_2$ *plus* Pa_3 & $Sa_3 \rightarrow Ma_3$ *plus* a Totality Fact (ensuring that a_1, a_2, and a_3 exhaust the domain) collectively ground and necessitate (by the standard logic of grounding), the overall regularity $\forall x(S\&Px \rightarrow Mx)$.

As before, this is a complex pattern of explanations, one which takes into account various kinds of constitutive and non-constitutive explanations, and which requires a vast array of *explanans* to function, from properties and dispositions, to vacuous L-instances of regularities, and even a Totality Fact. But it is independently plausible; furthermore, again, assuming that all such explanations entail necessitation, it allows us to reach the desired non-Humean conclusion that once the properties, and thus the dispositions, of the relevant objects are fixed, certain patterns of behaviour are also automatically fixed.

7.4.2 *Lack of a Unifying Explanation*

Let us now move to the objection to the account. The objection starts by observing a crucial difference between my account and other kinds of non-Humean accounts of governance: more precisely, this is a difference concerning how many items are required for the governing job. On my account, a different dispositional ascription is required to govern each non-vacuous L-instance of a regularity; and thus multiple dispositions contribute to necessitating a regularity. But on most other accounts, there is only one item doing the governing. Consider any kind of theory of governing laws, DTA or otherwise; as we have seen at the beginning of this chapter, for any regularity such that all Fs are Gs, this regularity is entirely determined and necessitated by the law that all Fs are Gs. Similarly, one can explain any individual also being G by virtue of the same thing: the law itself that all Fs are Gs. The explanation provided by the law is a *unifying* explanation, in the sense that the behaviour of multiple, potentially countless entities is ultimately traced back to one single item: the governing law. My pattern of governance, on the contrary, is rather scattershot: different dispositional ascriptions are responsible for different things behaving so and so, which is what ultimately the regularity consists into. Rather than a one-over-many governing item, we have a patchwork of governing items. The lack of a unifying element in the resulting governing explanations *could* of course be amended. The simplest way to do it would perhaps be to claim that there is something in common to all those dispositional ascriptions that warrant this kind of explanation, presumably because they are all instance of the same item, a unique disposition-type that is, by virtue of its very nature, particularly apt at making it the case that similarly disposed items behave in such-and-such a manner. The unifying element here would be the disposition-type, or perhaps its nature. This solution to the difficulty would surely bring us closer to Bird's (2007a) account of governance, in which a derivation of law-statements and regularities is obtained through the counterfactual-based essence of a potency – which is the same for each item instantiating the potency: soluble things dissolve in water because that behaviour is in the very nature of solubility itself (or the relevant potency).

Embracing this kind of solution means embarking on a substantial revision concerning the relata of governance, especially on the side of the *explanans*. That said, I must confess that I am reluctant to embark on such a project. As detailed in the introduction of this chapter, I would like to keep instances of governing as logically simple as possible, or more technically: I believe that a *ceteris paribus* an explanatory hypothesis on a sparser basis is to be preferred. In this case, I prefer an account of governance in terms of dispositional ascriptions, whose potentiality for perspicuity leads us into a discussion about the reality of dispositions without also embarking on a discussion on the reality of types, universals, essences thereof, and so forth. So the question is: is this problem sufficient to make us reconsider such a metaphysically deflated account of governance?

What is certain is that the emergence of this problem is not at all surprising given the current set-up; in fact, it was always a foreseeable consequence of my approach, or for any approach that tried to offer governance without postulating some superimposed governing item, be it a governing law or the essence of a unique property universal over-and-above the various things to be governed. Thus, one may claim that with the lack of a unifying explanation I am paying the price for my deflationism. I disagree to the extent that I am not entirely sure what price I am paying that is so unacceptable. According to the present reconstruction, every thing behaves in such-and-such a way because each one of them is so-and-so disposed: why would a unifying explanation be required at this stage? Of course, two further questions may be asked. First question: why are certain things behaving in the same way? My answer here would be: because they are similarly disposed (and, ultimately, because they are similarly propertied). I leave the details of this intuitive reply to another time. Second question: why is it that being so-and-so disposed explains why things behave in such and such a way? This is an important question, in which essentialism – and its promises of unifying explanations – may raise its head again: is there something about being so-and-so disposed and propertied, which is invariant amongst similarly disposed and propertied things, that ultimately explains why all such things behave the way they do? This is an intriguing meta-explanatory question that will be the main topic of the next chapter. For the time being, I stand by my claim that the non-Humean "engine" regulating the evolution of processes of certain entities is to be found within each one of such entities, and thus different each time.

7.5 Explaining Laws with Dispositions

Now that we have seen how dispositions can govern regularities, we can finally say something about how dispositions explain laws of nature.

Within my account, the reality of laws relates to the perspicuity of law-statements; to be clear, not simply the perspicuity of their content (which would amount to the joint-carvingness of expressions figuring in it, such as e.g., charge, distance, and the like) but the perspicuity of the very statement that certain regularities are not accidental, but nomic. In other words, the question of reality for laws amounts to the following question: can, and should, the distinction between law-statements and regularity-statements, be formulated in metaphysically privileged conceptual machinery?

The simplest and more radical option here would be nomic primitivism as expressed in Carroll (1994) and Maudlin (2007), which is particularly suited to be expressed within my account. As Maudlin (2007, p. 18) explains,

> [m]y analysis of laws is no analysis at all. Rather, I suggest we accept laws as fundamental entities in our ontology. Or, speaking at the conceptual level, the concept of a law cannot be reduced to other more primitive notions.

Given my ontologically deflationary set-up, it is on the second sentence that we must focus. The nomic primitivist's hypothesis here is that a sentential "L:" law-making operator is explanatory primitive and thus joint-carving. Nomicity is thus understood to be a genuine aspect of reality.

A slightly different situation is that of DTA, in which some degree of analysis is present, in the sense that nomicity is better expressed not as a sentential operator, but as a law-making second-order "nomic necessitation" binary relation N. That, presumably, is the primitive and thus a candidate for joint-carvingness. To the extent that the difference between these two positions is now reframed as a difference concerning which piece of conceptual machinery is metaphysically privileged, Hildebrand's (2013, p. 2) claim is vindicated that.

> we can distinguish [nomic primitivism] and [DTA] by their treatment of *statements* of laws. In its general form, [nomic primitivism] holds that the locution 'It is a law that' has a wholly primitive status corresponding to the ontological primitives capable of satisfying it. According to [DTA] the locution can be translated into different statements containing different primitive terms.

This point is, of course, merely exemplative: nomic primitivism and DTA are not the only options when it comes to the explanation (or lack thereof) of laws of nature. What is, then, a supporter of Dispositional Reality to do? On the present account, accounting for laws of nature through dispositions does not consist in the grounding of laws of nature, nor in finding truth-makers for law statements: it rather consists in finding constitutive explanations of law-statements through dispositions to hopefully prove that such dispositional *explanans* are more perspicuous than law-statements. In that case, nomicity itself viz. talking through "L:" law-making operator would out to be a metaphysically obfuscating way of talking about Dispositional Reality.

A crucial thing to notice is that, unlike the governing of regularities, we are here patently dealing with constitutive explanation: we are not simply trying to decide what determines, in a modal or more-than-modal way, what counts as a law of nature; we are rather trying to discern what the reality of laws of nature ultimately amounts to. But how should such constitutive explanation of laws be formulated?

The idea is actually very simple: whatever dispositional item is doing the governing, that is what the reality of laws ultimately consists of. If some dispositional region of metaphysical structure is supposed to perform the governing task previously performed by laws of nature in a non-Humean metaphysical systems, then to constitutively explain laws of nature all we need is to look for such dispositional portions of metaphysical structure.

Let us use again our toy-world example with the non-accidental regularity such that $\forall x(S\&Px \rightarrow Mx)$, with the quantifier ranging over our three-things domain of $\{a_1, a_2, a_3\}$; suppose that we want to constitutively explain, in our world, that "L: $\forall x$ $(S\&Px \rightarrow Mx)$". The reader will remember that two dispositional ascriptions govern two L-instances of the regularity, and thus contribute to the necessitation of the regularity itself: $D_{S\&P\text{-}M}a_1$ and $D_{S\&P\text{-}M}a_2$. However, those ascriptions taken collectively are not sufficient to necessitate the law, and thus cannot constitutively explain it. This is because two other items are required. One: Totality. Two: the vacuous L-instance concerning a_3 (which is not governed by any disposition, since by stipulation $\sim D_{S\&P\text{-}M}a_3$). These two requirements illuminate two crucial points concerning laws of nature that our account must preserve. Point one: laws must be universal, or at least sufficiently general. And two: law-statements, just like regularities such as $\forall x(S\&Px \rightarrow Mx)$ can be vacuously true, either partially or totally (some, or all, of its L-instances are vacuous). In particular, law-statements can be totally vacuously true in a possible world in which the relevant properties are entirely uninstantiated. This brings us to the topic of the so-called "alien laws", viz. laws of nature regulating the behaviour of alien properties (viz. actually uninstantiated properties). E.g., in a possible world in which no electrically charged particle exists, Coulomb's Law is vacuously true. Just as there are qualitatively impoverished worlds, there could also be qualitatively enriched worlds, displaying actually uninstantiated properties, whose behaviour is regulated by alien laws: perhaps the most famous example in philosophical literature (Fine, 2002, p. 257) is the case of objects that instead of having mass have schmass, a supposedly alien property obeying an inverse cube law as opposed to an inverse square law. If one is a bit worried about the reliability of this fantastical scenario, and suspects that the inverse cube law regulating the behaviour of schmassy objects is not actually true (not even vacuously true), one must note that nomic vacuity is not a philosopher's artifice, and is easy enough to spot in scientific theories. That is to say, there are several scientific principles deserving of the name of "law" that appear to apply to very few, if any, cases at all; e.g. the first law of motion[24] or the ideal gas law[25]; idealization is famously a source of vacuity, and is widespread enough to lead some philosophers to conclude that all laws of physics are vacuously true, if true at all.[26] But we do not need to assume as much[27]; for our purposes, it suffices to say that this conception of laws as potentially vacuously true, either partially or totally, is usually combined with the claim that law-statements are metaphysically necessary, a common claim amongst friends of dispositions (Ellis, 2001; Bird, 2007a), and usually argued for through a loosely Kripkean argumentative strategy: what about a world in which

[24] Ayer (1956) and Carroll (1990, 1994).

[25] Chakravarrty (2007, p. 141–147).

[26] Famously, Cartwright (1983).

[27] Broad (1935) provides what is effectively a vast taxonomy of vacuous laws.

objects obey an inverse cube law, as opposed to an inverse square law? Well, that's a world with schmassy objects, as opposed to massy objects; in that world, our law is still true, albeit vacuously so.

Some reader may find this conception of laws as unsettlingly overabundant: if laws about alien properties are actually true, then this means that there are far more laws of nature than those we were initially willing to acknowledge, viz. those discovered during the course of our scientific enquiry. For we also have to countenance laws for properties that we will never encounter. This is especially problematic for supporters of DTA or governing laws in general, who require a heavyweight conception of laws of nature to function – a metaphysically suspect inflation perhaps to be understood along Platonic lines.[28] But this isn't needed. The way I see it, the vacuous truths of certain law-statements is a mere consequence of the fact that the world is, so to speak, always ready for any kind of interaction or process: there's no possible process, as unlikely as it might be given the kind of properties that are actually instantiated, that is not already decided to go one way or the other.

For our purposes, these two points concerning laws (one: their universality, two: their potential vacuity) suggest us that there is something deeply right about the association between nomicity and universal quantification; this, in turn, may tell us something about what is constitutively explaining laws of nature. Yet, as useful as it may be, this consideration is only part of the answer: when we try to express the content of law-statements more perspicuously, what is it exactly that we are quantifying over? And what exactly are we saying about them?

To answer these questions, remember our toy-example again. What is the peculiarity of the vacuous L-instance concerning a_3? Unlike the other two items, a_3 is not P; it does not have the relevant disposition either, which is exactly why it contributes to the regularity in a vacuous way. In other words: a_3 contributes to the regularity that all things which are P&S are also M, by not being P&S. The same thing happens, on a larger scale, when a law is totally as opposed to partially vacuous: all actually existing objects are contributing to the truth of the law-statement regulating the behaviour of schmassy objects: they do it by way of not being schmassy. The reader may again look with suspicion at these so-called "contributions"; but again, remember that we are talking about constitutive explanation here, and not confirmation. Cases like that of a_3 in our toy example are cases of vacuous contributions to the universally quantified truth; these are limiting cases, so to speak: in the case of totally vacuous laws, all contributions are vacuous contributions.

All in all, there is an important lesson to be learned here: what all three entities in our toy-example have in common is that, necessarily, if they are P then they are also $D_{S\&P\text{-}M}$. This suggests a dispositional *explanans* of the right logical form for law-statements, viz. something along the lines of:

[28] As first suggested in Tooley (1987, p. 669).

14) L: $\forall x(S\&Px \rightarrow Mx)$ because$_C$ $\Box\forall x(Px \rightarrow D_{S\&P\text{-}M}x)$

Informally: that it's a law that S&P things are M is constitutively explained by the fact that necessarily, everything that is P, is also $D_{S\&P\text{-}M}$; according to the simplistic interpretation of our toy-example: it's a law that soluble things dissolve when immersed in water because necessarily soluble things are disposed to dissolve when immersed in water. More generally: laws of nature are constitutively explained through the modally robust correlation between properties and dispositions.

The first thing to point out concerning (14) is that the box operator in the *explanans* (motivated by the existence of an aforementioned necessitation between properties and dispositions) is required to avoid trivialization in cases in which quantification is vacuously true. If we suppose that nothing is P, then it is the case that everything that is P is also $D_{S\&P\text{-}M}$; but, as it turns out, it is also the case that everything that is P is also $\sim D_{S\&P\text{-}M}$. That's just how material implication works. Turning material implication in a strict implication solves this difficulty: if the property P is necessarily related to the right disposition $D_{S\&P\text{-}M}$, then $\forall x$ $(Px \rightarrow D_{S\&P\text{-}M}x)$ will be true at all worlds, vacuously or not; but if it is not, then there will be at least some world in which something that is P is not $D_{S\&P\text{-}M}x$, thus making $\Box\forall x(Px \rightarrow D_{S\&P\text{-}M}x)$ false, and thus non-trivial.

More generally, the *explanans* preserves all the features of a law-statement that we have listed above: it is sufficiently general, if not outright universal, and it can be vacuously true; as a matter of fact, such a dispositional *explanans* is vacuously true in exactly the same circumstances in which the law is: when nothing is P. Laws of nature, so understood, are ultimately about the connection between properties and dispositions: soluble things dissolve in water; electrically charged things attract other electrically charged things with a force of a certain magnitude; and so forth. In fact, there is nothing inherently revolutionary in such an approach[29]; its significance rather lies in its faithfulness to a philosophically sensible conception of what a law of nature is supposed to be: law-statements are not descriptions or summaries of regularities, nor do laws consist in such regularities; and neither laws are higher-order relations between universals, nor they are *sui generis* entities – as other philosophers have submitted. Laws of nature are, so to speak, blueprints for behaviour. Or rather: they associate properties with dispositions for certain patterns of behaviour; given that dispositions govern such behaviour (as I hereby submit) it follows from this account that laws of nature simply associate things that are-so-and-so propertied with governing items that lead to such-and-such process. Therefore, the nomic modality that laws of nature bring into the world is the properly non-Humean necessitation between how a thing is, and how it is disposed to behave (Ellis, 2001, p. 44).

Some reader may feel somewhat disappointed by the principle laid out in (14), and more precisely, by the appearance of a box operator in its right-hand side. What

[29] E.g., according to Ellis (2001, p. 204), "[i]n its simplest form, a law of nature says that for all x, if x belongs to the natural kind A, then x will be intrinsically disposed to be involved in an event of the natural kind B in circumstances of the kind C."

an awfully long way to go, just to have natural modalities explained by a hitherto undisclosed necessitation! Claims like (14) only raise further questions: what kind of necessity are we dealing with here? Is it a primitive affair, thus sanctioning the presence of residual modality in our account of laws of nature? Or can we account for it in some other way?

On the face of it, there is something that we can say about this necessity. As a matter of fact, we have said it already, in this very book: such a necessitating relation between properties and dispositions is a modal consequence of the non-constitutive and yet objective explanation of dispositions through properties provided in the previous chapter, which is non *ad hoc* and independently motivated by the need to articulate the complex relation between, say, electric charge, and the various dispositional tracks usually associated with it. Not only this can assuage some of the worries above, but it can also tell us something about the kind of modality proper of laws of nature. First of all, this is not the kind of lower-order necessitation proper of dispositions and causal relations: in stating (14) we are not claiming that causes necessitate, or tend towards, their effects with any modal strength. And neither we are expressing the kind of necessitation that is proper of governance: we are not stating that such-and-such objects behave in such a way, but only that they are *disposed to* behave in such a way. The kind of necessity in discussion here is higher-order: what we are stating is that laws, as blueprints for behaviour, correlate properties of objects with dispositions that they have, that is to say, with a nomic and causal profile: what modal force is proper of that dispositional and causal profile, remains to be seen (for a very similar tripartite distinction between kinds of necessitation, see Barker and Smart (2012, p. 721–722)).[30]

Relatedly, we must keep in mind that the explanation employed in (14) is constitutive explanation, in that the law 'just is' that *explanans*. The dispositional *explanans* of the law does not by itself govern the regularity (dispositional ascriptions do, albeit indirectly). To be clear, the law is not reduced to any disposition, nor collection of dispositions; the story in (14) is slightly more complicated. And neither is the law reduced to the regularity itself; for L: $\forall x(S\&Px \rightarrow Mx)$ is not constitutively explained by $\forall x(S\&Px \rightarrow Mx)$. Yet in a way, law-statements on the present account are a mere summary, a description of a non-accidental regularity, viz. the necessary pattern in the association between properties and dispositions. The present

[30] Some further meta-explanatory question lingers in the background: why are things with certain properties disposed to behave in such-and-such a way? Why does the world display certain blueprints for behaviour, instead of others? This is a very broad question, and a difficult one to answer. In fact, I expect that a question like "why do *these* laws of nature hold, instead of others?" will receive a variety of different answers, depending on the kind of laws under discussion, and the kind of context-sensitive interests that would make an explanation successful. When it comes to metaphysics, meta-explanatory questions such as these often lead to talk of *essences*. Electrically charged particles, we are told, are disposed to attract and repel each other, because that is in their nature to do so; this is kind of answer that we can expect from positions such as Bird's dispositional essentialism. But even putting aside the kind of explanatory/determinative circularities often associated with such positions, one can feel an air of circularity and vacuity in this essentialism-back answers. I will come back to essentialism and meta-explanatory questions in the next chapter.

account of laws of nature is closer in spirit to the lawless account to be found in Mumford (2004). This is true in more ways than one: even if law-talk was to be perspicuous, and so, even if $\mathbb{S}$(L:), that wouldn't require the reification of laws-as-entities – albeit it would still require that laws are part of metaphysical structure in the sense that nomic facts objectively resemble each other. But that is still conceding too much to realism about laws: if the present account is correct, and the constitutive explanation in sentences like (14) is genuine, then those objective similarities would be better expressed by the more perspicuous talk in the right-hand side, which includes dispositional ideology. Law-statements are true, but there's no such things as heavy-duty governing laws. In the words of Ellis (2001, p. 128): "laws are not things that exist in the world; they are things that are true of the world".

7.6 Conclusions

The postulation of Dispositional Reality as a part of metaphysical structure is not meant to be a theoretically idle postulation: that things are so-and-so disposed is supposed to make a difference in the world. This is, first and foremost, a causal difference: things so-and-so disposed enter certain causal interactions and processes, instead of others. But it is not merely a causal difference; that very causal difference can itself be explained in terms of the disposition itself. As a matter of fact, the study of governance as we have pursued it in this chapter, is the study of how exactly positing dispositions makes a causal difference in the world; not simply in the sense that the dispositions function as causes (something that I haven't explicitly tackled), but because a thing being so-and-so disposed leads to certain causal interactions and processes, instead of others, and thus ultimately determines the distribution of properties across space-time.

This took the form of an explanatory hypothesis. I have investigated multiple kinds of explanations, and tried to position governance amongst them: governance ought to be understood as a *sui generis* non-scientific, non-causal, and non-constitutive explanation, perhaps to be understood, like causal explanation and unlike constitutive explanation, in a realist spirit. Furthermore, I have tried to address how to constitutively explain nomicity in terms of dispositions, thus reaching the tentative conclusion that even if a governing non-Humean metaphysics is our goal, there is no need to take nomic talk as metaphysically perspicuous.

These explanations have been pursued with a deflationary spirit in mind, discharging property-talk, essence-talk, and similar conceptual resources usually deployed by non-Humeans. Our *explanans* are comparatively simpler; the goal being, again, to lay the foundation of a theory of Dispositional Reality with some winning theoretical virtues, such as simplicity, number of primitives, and so forth. This chapter thus functions as partial corroboration for Dispositional Reality as an explanatory hypothesis.

Chapter 8
Meta-Explanatory Questions

8.1 Introduction

Although its supporters do not always explicitly state it, DTA entails the modal principle

1) $\Box(N(F,G) \rightarrow \forall x(Fx \rightarrow Gx))$.

Yet this is far from the only position in which a strict entailment exists between laws of nature and regularities. As in Schaffer (2016a, p. 578), the same problem exists for the nomic primitivist position in Carroll (1994) and Maudlin (2007), in which nomicity is expressed by a primitive "L:" law-making operator; to the extent in which governance, as an explanatory relation, entails necessitation (and to the extent in which we decide to use the orthodox formalization of law-statements and regularity-statements), we may formulate the following modal principle:

2) $\Box(L{:}(\forall x(Fx \rightarrow Gx)) \rightarrow \forall x(Fx \rightarrow Gx))$.[1]

In both (1) and (2) laws are taken to be distinct existences from regularities. This can be easily expressed through our generalized identity operator:

3) ~(N(F,G) *just is* ∀x(Fx → Gx)),

which makes (1) a non-Humean necessary connection between distinct existences (and thus the correspondent kind of explanation non-constitutive). And similarly:

[1] Nomic primitivists do not always claim that laws govern regularities. But given the characterization of governance provided in the previous chapter, there is at least enough evidence for the compatibility between nomic primitivism and governance. Governance is a peculiarly non-Humean form of non-scientific explanation, and nomic primitivists in fact claim that laws explain regularities in a way that is unavailable for neo-Humeans, as Carroll (1994, p. 108–109; 2008, p. 76–77) and Maudlin (2007, p. 178). Of course, this only shows that governance and nomic primitivism are compatible, which is enough for our current purpose.

L. Azzano, *Dispositional Reality*, Synthese Library 482,
https://doi.org/10.1007/978-3-031-52625-1_8

4) $\sim(L{:}(\forall x(Fx \rightarrow Gx)$ *just is* $\forall x(Fx \rightarrow Gx))$,

which makes (2) a non-Humean necessary connection between distinct existences (and the correspondent kind of explanation non-constitutive).

The question thus emerges: whence the necessity in (1) and (2)? The problem is not simply that these necessities between distinct existences violate some highfaluting neo-Humean metaphysical dogma. Even from a non-Humean perspective, the problem remains as to whether we should just accept brute necessary connections, or explain how these necessities come to be.[2] As already suggested, identity can explain a necessity: necessarily, Superman and Clark Kent are always co-located, which is because they are identical. But identity is not the only relation that can explain an identity, and given (3) and (4), we must clearly seek for another *explanans*. So, what relation is there between laws and regularities to explain modal principles such as (1) and (2)? The answer that we have provided in the previous chapter is: governance. Governance is a *sui generis* relation of non-causal explanation, realist in the sense that is underpinned by (or perhaps identical to) a relation of non-causal determination which is more-than-modal but at least modal – thus entailing, and ultimately motivating, modal principles such as (1) and (2). But this only defers the question: how exactly do laws govern regularities? This is a meta-explanatory question, in the sense that it seeks to explain an instance of governance. Given our characterization of governance, it seeks to clarify an explanatory connection between laws and regularities; furthermore, given that it is a modal-cum-explanatory connection that we are trying to clarify, the question becomes: how do laws govern, and thus necessitate regularities? In other words: how exactly are the modal restrictions of property-distributions in the mosaic generated by laws, dispositions, or what have you?

Meta-explanatory questions concerning (1) were first asked in Lewis (1983) and van Fraassen (1989), and replicated for (2) in Schaffer (2016a). These criticisms are not usually presented as meta-explanatory challenges, but as inference challenges, viz. as requests to justify the entailment, or strict entailment, between laws and regularities. But it seems to me that these "inference challenges" are but a steppingstone towards meta-explanatory challenges. Consider an analogy with constitutive explanation: suppose that being crimson strictly entails being red. Why is that? The answer may be that being crimson constitutively explains being red, and thus necessitates it. But why exactly is it that being crimson constitutively explains being red? The inference challenge has been answered, but the meta-explanatory question lingers. The same is true for governance: the (strict) entailment between laws and regularities may be explained by the presence of an instance of governance: but why, or how exactly, this governance between the two items

[2] For the possibility of brute necessities, see Barker and Smart (2012) and Barker (2013).

occurs? The meta-explanatory question remains.[3] In conclusion, we can take these "inference challenges" to ultimately amount to meta-explanatory challenges.

The challenge is not the same for everyone, albeit there is a general pattern that we can follow. Of course one may think that DTA has it better than nomic primitivism in this regard, as non-primitivism about laws can at least attempt to articulate what exactly it is about laws that ensure regularities.[4] But similar meta-explanatory problems patently arise for any position in which laws of nature are distinct existences from the regularities they necessitate, and, more generally, for every non-Humean position in which governance is involved. This is patently true for many forms of dispositionalists, e.g. for Bird's (2007a) position in which relational essences of properties govern and necessitate regularities, which, as brilliantly demonstrated by Barker and Smart (2012), suffers from the same problem than Armstrong's variant of DTA did, the only difference being whether second-order relations between universals are part of the essence of properties or not: in other words, "[Bird's] dispositional monism is basically Armstrongian necessitarianism modified to allow for a thesis about property identity" (Barker & Smart, 2012, p. 714). After all, on Bird's account it is still the case that, given the essential second-order SM relation between stimuli (S) and properties (P) on the one side, and manifestations (M) on the other:

5) $\Box(SM((S,P),M) \rightarrow \forall x(Sx \& Px \rightarrow Mx))$.

I won't enter these specific debates here.[5] But the issue is arguably generalizable, so the following doubt may creep in: is my take on Dispositional Reality and governance subject to these meta-explanatory questions? The answer is: yes, which motivates the present chapter.

I previously discussed how certain dispositional ascriptions may govern (and thus necessitate) non-vacuous L-instances of regularities, such that, e.g.:

6) $Fa_1 \rightarrow Ga_1$ because$_G$ $D_{F\text{-}G}a_1$

And thus:

[3] Incidentally, an analogous point can be made concerning causal explanation, which invites the question: can there be meta-explanatory questions concerning causal explanation? The case might be slightly different here, as the modal strength of causation is plausibly not that of strict entailment, but a similar point can be made as well: how can the modal correlation between, say, smoke and fire be motivated? This inference question can be answered by claiming that smoke is causally explained by fire, which leaves the door open for a meta-explanatory question: how does fire explain smoke? As we will see throughout this chapter, the comparison with causal explanation will be particularly useful in understanding how to answer such meta-explanatory questions.

[4] For one, Hildebrand (2013, p. 5) argues that a primitive sentential nomic operator such as "L:" offers little reason to think that it must always accompanied by statements of uniform regularities: whatever these *sui generis* laws are, they do not in fact ensure regularities.

[5] But see Tugby (2012) for a reply to Barker and Smart (2012) on behalf of the friend of dispositions. I will come back to this (in my opinion unsatisfactory) strategy in Sect. 8.4.

7) $\Box(D_{F\text{-}G}a_1 \rightarrow (Fa_1 \rightarrow Ga_1))$.

The pair (6)–(7) openly invites meta-explanatory questions. Is the association between dispositional ascriptions and L-instances primitive, or can it be analysed any further? As we have seen, Armstrong tried to pursue the claim that there was something about the nature of N, as a supposedly "nomic necessitation" relation, to explain how laws could play their governing role; can a similar strategy be attempted here? Is there something about the nature of dispositionality that brings about certain property patterns as opposed to others? Understanding how exactly the governance operated by dispositions works may *prima facie* strike like a request for an underlying mechanism regulating dispositionality as a modal/nomic primitive of the theory – a request to "dissect dispositionality" so to speak – but whether that is possible, or even desirable, remains to be seen.

What is more, this pattern of meta-explanatory questions can be replicated beyond governance, to other explanatory endeavours discussed in this book, for example:

(a) the non-constitutive explanation between properties and dispositions presented in Chap. 6.
(b) the constitutive explanation between laws and dispositions presented in Chap. 7.

The possibility of meta-explanatory questions constitutes a common worry concerning of all these explanatory endeavours.

There's much at stake here: most importantly, the status of dispositionality as an explanatory primitive. By this point of the book, we have a clear grasp on Dispositional Reality and the kind of metaphysical heavy-lifting that it is supposed to do, as an explanatory hypothesis, regarding laws, regularities, and the like. By investigate meta-explanatory questions, we can therefore ask the important question: is dispositionality a primitive of the account? Or is there something else ultimately performing explanatory jobs? We will tackle this issue by drawing not only from the debate on the "inference challenges" seen above, but on recent debates concerning meta-grounding questions. Although we must keep in mind that governance is not constitutive explanation, there are important lessons from that debate we must heed. The primitivist solution I will present, although explicitly aimed to deal with meta-explanatory questions concerning dispositions, may in fact generalize to other kinds of meta-explanatory endeavours, such as the one concerning DTA and nomic primitivism, as above; but this falls squarely beyond the scope of this book.

In Sect. 8.2 I will briefly introduce the debate on meta-grounding questions, and in Sect. 8.3 I will discuss its fallout within the framework put forward in this book. The main takeaway of these sections, explored in Sect. 8.4, is that meta-explanatory questions concerning disposition-based explanations should be answered by providing meta-explanatory links connecting the two ends of the relation; since there are different kinds of explanations at play here, different kinds of meta-explanatory links will be needed. Such meta-explanatory links are to be understood as postulational posits, introducing a metaphysically loaded notion of dispositionality within the theory. Another crucial element discussed in Sect. 8.4 concerns the presence of essence-talk in these meta-explanatory answers: although it is a common claim amongst friends of dispositions that things behave the way they do because of the *nature* of the dispositional properties they possess, I go on to argue that this essentialist component (even when properly reframed within the context of Dispositional Reality) is in fact explanatorily inert. Finally, in Sect. 8.5, I will clarify that certain dispositional ascriptions are explanatorily fundamental, and thus that there is no way to "dissect dispositionality": yet this is conclusion does not trivialize the explanations provided throughout this book.

8.2 Meta-Grounding Questions

It will be highly instructive to begin from the discussion concerning meta-grounding questions, viz. questions about what grounds grounding facts. This may strike the reader as a detour, but it is the most straightforward entry-point to the discussion on meta-explanatory questions; as per Chap. 3, I do not employ a standard notion of metaphysical grounding as my go-to tool in meta-metaphysics, although I do have a proxy notion in constitutive explanation that plays some of the roles standardly played by metaphysical grounding (although not all of them). In the next section I will discuss the fallout of the meta-grounding debate within my account developed around the notions of metaphysical perspicuity and non-ontic metaphysical structure. But for now, let us introduce the debate on meta-grounding questions.

8.2.1 Meta-Grounding, Purity, and Collapse

Here's a familiar picture: suppose that there is a binary relation of metaphysical grounding, layering the facts of reality along a hierarchy of relative fundamental; only at the bottom, if there is one, do we find what is absolutely fundamental (or fundamental for short). If one feels queasy about ontological commitment to facts, one can simply use a sentential operator $<$ to express grounding, such that, e.g.:

8) The ball is crimson $<$ The ball is coloured.

Assuming that for no sentence ϕ, it is the case that $\phi <$ The ball is crimson, we can then say that the ball being crimson is amongst the fundamental facts of the world, viz. it is fundamentally the case that the ball is crimson. To use the standard God-metaphor: when God was creating the world, They did have to make the ball crimson, but They did not have to make it coloured: that came for free when They made it crimson. Now, what about (8) itself? Is there some ϕ, such that $\phi <$ (8)? If so, what could that meta-grounding fact be? This is a question about the fundamentality, and thus the reality, of grounding facts: when God was creating the world, did They have to create grounding as well?

Before we proceed, there is a vital clarification to make. The question is not whether the relation of grounding, *by itself*, is fundamental, or real. This question (which has we have seen in Chap. 3, on my account in cannot be side-stepped, nor answered one way or the other simply by choosing between predicative vs. operationalist accounts of metaphysical grounding) cannot even be meaningfully asked on this setting, since what is fundamental is what is fundamentally the case, in the sense that its contents need to be expressible sententially; such that, e.g. it is fundamentally the case that ϕ if for no Ψ, $\Psi < \phi$. Here, as before, I'm using metalinguistic variables to avoid talk of facts; but we could also say, if we are not worried about talk of facts, that to be fundamental are facts, and the relation of grounding is not a fact. So there is no asking about whether grounding is fundamental – although there questions in the vicinity of this that may be answered, as in Litland (2020, p. 144–145).[6]

The question here is whether grounding facts are fundamental. This seems like a straightforward thing to ask, a question made possible by the logical form of grounding, and avoided only at the cost of imposing a potentially *ad hoc* restriction on the kind of value that the metavariables Ψ and ϕ, as above, might receive.[7] All in all, meta-grounding questions generate "an awkward dilemma at [the] foundation" (Sider, 2020, p. 748) of the theory of metaphysical grounding. Why exactly is the question about what grounds grounding facts "a dilemma", in Sider's words? Part of the reason is that, based on certain assumptions of the nature of fundamentality, there are reasons to think that grounding facts can never be fundamental. This is the so-called Argument from Purity. The intuition behind the argument is that in fundamental facts we can only find fundamental objects and properties or relations; or perhaps that in fundamental truths we can only find fundamental notions; since this assumption of purity was first expressed in Sider (2011), it is particularly suited to be expressed within the lenses of my approach as well: if we take grounding sentences to be fundamental in the sense that they are metaphysically perspicuous then this entail that all sub-sentential components, in both sides of the grounding claim, are joint-carving. Consider for instance sentence (8): if it was fundamental, or perspicuous, then not only would "being crimson" a fundamental notion, as perhaps

[6] Schaffer (2009) champions a position according to which metaphysical grounding as a trans-categorical relation, which may also relate objects. Although grounding is usually regimented as either a relation between facts, or a sentential operator, one will occasionally find in the literature claims that grounding relates objects. If that is the case, it may be possible to directly discuss the fundamentality of the grounding relation qua object, by using "$<$" as a name for that relation. I will not say more on this matter. On my account, the fundamentality or reality of grounding is conducted in another way, viz. by discussing the perspicuity of sentences like (8), and the joint-carvingness of the $<$ operator; so the resolution of this alternative method is not terribly important for our present purposes.

[7] As we will see shortly, some philosophers do in fact claim that such a restriction ought to be put in place to stop meta-grounding questions from coming; although how exactly should this restriction be formulated, remains to be seen.

intended, but "being coloured" as well: after all, that expression also appears in the grounding claim. This problem generalizes, and can be expressed in a variety of different ways; if, say, facts about cities are grounded in more fundamental facts about its components, then to claim that this grounding fact is fundamental would also entail that cities themselves are fundamental – since cities appear in one side of the grounding claim. Yet the very point of the grounding claim was to establish that cities weren't fundamental, viz. that when God was creating the world arranging its fundamental objects, whichever they were, cities and facts about cities came for free.

From this kind of observation, we can obtain a veritable Collapse Argument (2013) against the claim that grounding claims are fundamental. Litland (2020, p. 134) offers a clear formulation of the argument:

> an object *b* is fundamental$_o$ iff *b* is a constituent of a fundamental fact. [...] The Collapse Problem is that if grounding facts are ungrounded, all objects *b* are fundamental$_o$. Consider some fact involving *b*, e.g., the fact that *b* exists (call it *Eb*). Either *Eb* is ungrounded or it isn't. If it is, *b* is fundamental$_o$. If it isn't, there is some such that $\Gamma < Eb$. But since the fact that $\Gamma < Eb$ is ungrounded, *b* is fundamental$_o$ after all.

We can also formulate a Collapse Argument using a set-up that more closely resembles the one employed throughout this book. In other words, we may try to expand on the idea of "fundamental notions" or "fundamental concepts" (Sider, 2011, 2020), by claiming that an expression α of arbitrary syntactic category is fundamental$_e$ if it appears, at least once, in a sentence ϕ such that for no sentence Ψ, $\Psi < \phi$. A fundamental$_e$ expression α may be a predicate, such as the predicate "is crimson" in (8); but as our preferred notion of joint-carvingness, it generalizes to other kinds of expressions: referential terms, sentential operators, quantifiers and so forth.

Simply put, if an expression α appears in at least one fundamental (viz. ungrounded) sentence, then it is a fundamental$_e$ expression. Purity now takes the form of the claim that in fundamental sentences, there only appear fundamental$_e$ expressions. This is quite a simple thought, really: it ultimately amounts to the claim that in ungrounded sentences, there only are expressions that appear at least once in an ungrounded sentence – as a matter of fact, that very sentence. But once we assume that grounding sentences are ungrounded, there are disastrous consequences. For any expression α and any sentence ϕ in which α appears, it is the case that either ϕ is fundamental, or it isn't. If it is, then α is fundamental$_e$, by definition. If ϕ is not fundamental, then there is some other sentence Ψ such that $\Psi < \phi$; if α appears in ϕ, then α also appear in the grounding sentence $\Psi < \phi$, and if grounding sentences are fundamental, then $\Psi < \phi$ also is fundamental; so, by definition again, α is fundamental$_e$. Because in this two-horned dilemma we did not make any assumption regarding this arbitrary expression α (except that it appears in at least one sentence), this means that all expressions are fundamental$_e$. Thus, the Collapse. If grounding is fundamental, everything is.[8]

I will not question the validity nor soundness of the Collapse Argument, nor the Purity assumption that underlies it. Although there surely are ways to circumvent them, both are in my mind quite compelling. Neither I will accept the conclusion of the argument, viz. that everything is fundamental: for metaphysical grounding is explicitly deployed to make sense of the idea that *not* everything is fundamental. For the moment, I will draw the intended conclusion that grounding claims are not ungrounded, which naturally raises the further question: but if meta-grounding questions can receive non-trivial answers, which kind of answer should we expect? This is what we now turn to discuss.

8.2.2 Answers to Meta-Grounding Questions

Perhaps the simplest systematic answer to meta-grounding questions,[9] as in Bennett (2011), deRosset (2013), and Litland (2017) is what we may call the Straightforward Solution – but also see Bennett (2017) for an essentialist version of this strategy. Shortly put, according to the Straightforward Solution, for every ϕ and Ψ, such that.

9) $\Phi < \Psi$,

it is also the case that.

10) $\phi < (\phi < \Psi)$.

Informally, e.g., if the fact that the ball is crimson grounds the fact that the ball is coloured, then the fact that the ball is crimson grounds the fact that the ball is crimson grounds the fact that the ball is coloured. This unlovely mouth ultimately amounts to

---

[8] We could course-correct our definition by claiming that fundamental$_e$ expressions *only* appear in ungrounded sentences. But clearly, that would set the bar too high: identity might be a fundamental relation, but it does not only relate fundamental objects; electric charge might be a fundamental property, but it is not only possessed by fundamental objects; modality as expressed by standard sentential operator might be fundamental, but there are modal facts about non-fundamental objects and properties as well. More to the point, together with previous assumptions (Purity, and the claim that grounding claims are fundamental), this correction would generate a contradiction. Here is how: Because of Purity, a sentence is non-fundamental if at least one non-fundamental$_e$ expression appears in it. But if a non-fundamental$_e$ expression is one (as hereby proposed) which appears in at least one non-fundamental sentence, this ultimately amounts to the claim that fundamental sentences only contain expressions that only appear in fundamental sentences. Consider an instance of grounding $\Psi < \phi$, which is hereby conceived as fundamental; any expression α that appears in $\Psi < \phi$ also appear in ϕ. By what has been said above, if $\Psi < \phi$ is fundamental, and thus α is fundamental$_e$, any sentence in which α appears is also fundamental; thus ϕ also is fundamental: and yet ϕ is *not* fundamental, by our initial stipulation that $\Psi < \phi$.

[9] For non-systematic solutions, see Sider (2020). This distinction is not terribly important for the purpose of this chapter.

the idea that the ball being crimson makes itself suitable to ground the ball being coloured.

As one can clearly see by sentences (9)–(10), the Straightforward Solution is recursive: given (9)–(10), it is also the case that:

11) $\phi < (\phi < (\phi < \Psi))$
12) $\phi < (\phi < (\phi < (\phi < \Psi)))$
13) etc...

This may seem *prima facie* absurd, but it confers a twofold advantage to the Straightforward Solution. Firstly, it allows one to answer *all* meta-grounding questions. This is no small feat: a feature of the debate on meta-grounding questions – and a crucial one for the purpose of this chapter– is that meta-grounding questions can always be formulated: for whichever ground for a grounding claim one provides, the question can always be asked: what grounds that meta-grounding claim? The Straightforward Solution answers all of these questions in a single stroke; it is not a theoretically costly move, either: as noted in Bennett (2017, p. 197), this doesn't violate well-foundedness (it is one thing to claim that there are infinite chains of grounding claims, and another thing to claim that there is an infinite series of nested grounding claims, as in (9)–(12)). Secondly, it avoids the Collapse Argument and all considerations from Purity. The Straightforward Solution easily sidesteps this difficulty: no fundamental (viz. ungrounded) sentence is ever reached, and thus Purity never enters the picture; furthermore, the sentence recursively appearing the grounding base ϕ is expressly one in which the supposedly non-fundamental notions do not appear.

As fascinating as it is, however, there is a main difficulty plaguing the Straightforward Solution; simply put, the problem is that the answers it provides are not very convincing. According to Dasgupta (2014b, p. 752):

> the proposed explanations sound bad. To see this let us consider one in the flesh. Question 1: Why is this event a conference (rather than, say, a football match)? Answer: Because it contains people engaged in C-activities. So far, so good. Question 2: Why is it that those activities make the event count as a conference (rather than a football match)? The simple reductionist says: Because those people engaged in C-activities. This is not a good explanation.

This may strike as a rather vague observation, but it actually points to a crucial issue concerning the very nature of grounding *qua* explanation, an observation which straightforwardly carries over to my account. The reason why meta-grounding answers such as the one above do not appear to be informative, is because they lack any general principle connecting the notions from the two sides of the original grounding claim; to simply reiterate that the ball is crimson does not answer why the fact that the ball is crimson grounds the fact that the ball is coloured: a more general principle connecting being crimson and being coloured is needed. As for the nature and features of this connecting principle, I will say something shortly. But this is a feature of explanation that generalizes beyond the kind of "metaphysical explanation" usually connected to grounding. The standard DN model for scientific explanation, for example, couldn't function without a covering law allowing the deduction; and similarly causal explanation couldn't function without a causal law of nature connecting cause and effect in the appropriate way. Without such connecting principles, no understanding, and thus no explanatory knowledge would be achieved concerning the *explanandum*, given only the *explanans*. Relatedly, Schaffer (2017, p. 309) offers the following generalization:

> [i]nsofar as causal and metaphysical explanation are both forms of explanation, it becomes possible to use the structure of one to gain insight into the structure of the other. Insofar as causal explanation requires laws of nature (and overall involves a <*Sources*, *Links*, *Result*> dependence structure), metaphysical explanation has a structurally parallel requirement. Without linking principles, nothing connects the sources to the result, no general pattern of dependence is revealed, and a full understanding of why the result obtains remains elusive, in causal and metaphysical cases equally.

Mind you, the passage above does not imply that grounding-based explanations and causal explanations are two species of the same genus, or that grounding is "metaphysical causation". It relies on a weaker claim: that they are both explanatory. The need of connective principles is not simply a requirement for causal explanation, whether that includes grounding: it is requirement for explanation in general.

So, if grounding *qua* explanatory relation (along unionist or separatist lines), then there are genuine connecting principles which may provide answers to meta-grounding questions. We may call Connectivism (Dasgupta, 2014b) a family of solutions to the meta-grounding issue which follows this general structure. These connective principles may come in different flavours, thus generating different variants of Connectivism. Borrowing from Dasgupta (2014b, p. 568), the Connectivist may answer the meta-grounding question concerning (8) in the following ways:

13) It is essential to anything that is crimson, that it is coloured
14) It is (metaphysically?) necessary that if something is crimson, then it is coloured
15) It's a conceptual truth that if something is crimson, then it is coloured
16) It's a metaphysical law that if something is crimson, then it is coloured

As different as they are, there is a crucial issue that is relevant to all these variants of Connectivism: that is to say, how are they going to stop the meta-grounding regress – which was so effectively dealt with within the framework of the Straightforward Solution. The simplest solution is simply to claim that sentences like (13)–(16) are in fact ungrounded, which distinguishes another family of Connectivist solutions, which we may call Brute Connectivism, as in Dasgupta (2014b, p. 568). Of course, the problem with this approach lies in the fact that "is coloured" appears in these allegedly fundamental (13)–(16), so it would seem that, unlike the Straightforward Solution, this kind of answer to meta-grounding questions only postpones the problem, rather than solving it.

The standard reply to this argument against Brute Connectivism lies in the claim that these meta-grounding principles are a special kind of ungrounded fact: they are not, so to speak, within the fundamentality hierarchy of basic vs. derivate fact, and thus consideration of Purity and fundamentality do not even arise. Meta-grounding connecting principles are "not apt for being grounded in the first place, by which I mean that the question of what grounds it does not legitimately arise" (Dasgupta,

2014b, p. 576). This is the position called Trialism (deRosset, 2013; Litland, 2020). Of course the supporter of Trialism must elaborate and explain exactly why these meta-grounding principles are special in this way – why they are not apt for being grounded. There are various ways to cash out this notion, but they mostly revolve around the idea – perhaps the driving force of Trialism in the first place – that to seek grounds for connecting principles like (13)–(16) is something akin to a category mistake (Dasgupta, 2014b, p. 576); if Schaffer's (2017) aforementioned account of the structure of explanation as involving a triple <*Sources, Links, Result* > is understood as providing some insight on the logical form of explanation, then the quest for the grounds of (13)–(16) would be as misplaced as a quest for the causal explanation of a causal law of nature. But there are other ways to justify the distinctions posited by Trialism, mostly revolving around the idea that the meta-grounding principles of Connectivism are not revolve around "worldly facts" for which it makes sense to ask grounding questions, but rather constitute what we may call an explanatory infrastructure which makes grounding possible in the first place.[10]

These are complicated matters to assess. It may be in fact impossible to completely evaluate Brute Connectivism at this level of generality; but it is not the present objective either to resolve standard meta-grounding issues. Instead, we must make things a little less abstract, and try to understand how this entire meta-grounding issue should be understood within my metaphysical framework – and how it translates into a request for meta-explanatory answers as discussed in the introduction to this chapter. Only then a solution will emerge.[11]

8.3 Meta-Grounding Recontextualized

What remains of meta-grounding questions within my account? And how does that relate to the meta-explanatory questions I asked in the beginning of the chapter?

The main point of contention here is that, as per Chap. 3, I do not really employ metaphysical grounding as a layering of reality. So there is a reading of the situation that makes it moot vis-à-vis my account. Yet this needn't need be: a version of the Collapse Argument in fact remains, as we will see, although my account straightforwardly offers a way out.

When setting up the Collapse Argument, deRosset (2013, p. 1) paints a very general albeit familiar picture: "there is a layered structure of facts and the entities

[10] See Litland's (2020, p. 142) claims concerning Dasgupta (2014b, p. 562).

[11] Finally, there is a further systematic way to answer meta-grounding questions so as to avoid the Collapse Argument, which also exploits a distinction amongst the fundamental (or ungrounded) facts. I am talking about Fine's (2012) distinction between *ungroundedness* and *zero-groundedness* (viz. being grounded by the empty set of facts or sentences), which allows one to claim, as in Litland (2017), that although grounding facts are not grounded by any specific fact, they are zero-grounded instead of ungrounded, and thus are not, strictly speaking, fundamental. Also see Wallner (2021).

those facts involve". The simplest interpretation a grounding theorist could propose of this observation is the most literal one: there is an ordering of relative fundamentality of facts, and perhaps objects too, imposed by a standard relation of metaphysical grounding as a strict partial order. At the bottom, we find what is fundamental *tout court*. Within this context, it is easy to see how the problem arises: a well-formed grounding sentence such as (8) should express a fact, just like the others. We lack any reason to think otherwise. So the question immediately arises as to where, in this fundamentality structure, we should find grounding facts; then considerations of Purity arise, and the Collapse Argument can be formulated.

It is all very literal. It is also not what I believe to be the case. The simplest reason is that I do not believe there to an out-there relation of metaphysical grounding layering reality. This is not simply achieved by switching away from a predicative stance which requires heavy-duty facts or objects to be discussed as either "fundamental" or "derivative". For I do not simply reject layers: I reject the layering itself, a metaphysical stance that cannot be expressed merely by switching from a predicative to an operationalist stance. Now, there *is* a notion of constitutive explanation at the centre of my methodology for metaphysics, yet that notion is understood along irrealist lines: it is not part of the furniture of the world, but merely an explanatory hypothesis about what is more perspicuous. This block considerations of Purity from arising in the first place. Consider the question:

17) The ball is coloured because$_C$ the ball is crimson

On my account, a Collapse Argument concerning fundamentality would be formulated by claiming, by virtue of Purity, that in an (absolutely) perspicuous sentence, all of its sub-sentential components are joint-carving. So if (17) is (absolutely) perspicuous, then $\mathbb{S}$(is coloured); hence, the Collapse. But I am not particularly worried of this possibility: (17) is patently *not* metaphysically perspicuous, because ~$\mathbb{S}$(because$_C$). Thus, enter meta-explanations: we may formulate a question of constitutive explanation concerning explanatory claims such as (17). This is not just the case for constitutive explanation; for any kind of explanation because$_X$, and any arbitrary sentences ϕ and Ψ such that.

18) ϕ because$_X$ Ψ,

We can legitimately wonder how to provide completions to:

19) (ϕ because$_X$ Ψ) because$_C$ …

In other words: what does the explanatoriness of Ψ vis-à-vis ϕ consist into? Depending on the kind of explanation because$_X$ we are dealing with, different answers will be provided. Now, it is important to point out that answering these kinds of meta-explanatory questions does not amount to defining the relevant explanatory notions: for whether, and how, can explanation be defined is a long-standing issue in philosophy. But to ground is not the same as to define (Sider, 2020), and the same is true for constitutive explanation. So we can try to offer completions in sentences like (19) to offer some crucial information as to why certain claims are explanatory in the relevant sense, without aiming to give a general definition of explanation.

Based on the primarily epistemic notion of explanation employed so far, it is the primary job of an explanation to provide understanding (this notion is occasionally circularly cashed out through the notion of explanatory knowledge – but again, definition is no matter here). I agree with the point made by the Connectivism that no understanding is possible without grasping some kind of link between *explanans* and *explanandum*, as the explanation itself is meant to relate the two within our epistemic corpus (Kim, 2010, p. 150). But whether or not any provided link will be explanatory depends on the context of the explanatory enquiry, and the nature of the link itself. For example, consider Maria, who is struggling to learn English and doesn't understand the word "bachelor"; she will probably not understand the supposedly explanatory claim:

20) Tim is a bachelor because Tim is an unmarried man

Maria struggles with (20) in the sense that she struggles to understand what the explanatoriness of Tim being an unmarried man consists into vis-à-vis Tim being a bachelor. A certain kind of link can be provided:

21) It is the definition of the concept "bachelor" that to be a bachelor is to be an unmarried man

Hence, the confusion is dispelled. There is an important lesson to be learned here. Before hearing (21), Maria was very much confused about the nature of the supposed explanation in (20); the kind of explanation hereby involved becomes clear only after hearing (21). Is Tim being an unmarried man a causal explanation of Tim being a bachelor? A scientific explanation? A "metaphysical" explanation (whatever that means)? Or a constitutive explanation? This only becomes clear after hearing that the meta-explanatory link is a conceptual definition, which makes it clear that the explanation involved in (20) approximates some kind of conceptual explanation. In other cases, things are different; e.g. Maria might be confused by the supposedly explanatory claim.

22) The window shattered because Tim threw a rock at it

She can then be provided a concatenation of causal laws linking the two events, which will help Maria understand that the supposed explanation in (22) is a causal/ scientific explanation. In short, the nature of the meta-explanatory link decides the kind of explanation at hand (whether the overall *explanans* should consist in the <Source, Link> pair, or if explanation should irreducibly be understood as triadic, is a bookkeeping issue concerning logical form I am not particularly interested in discussing). Sometimes, Brute Connectivism seems like the right option; this is the case of (20): after hearing that the concept of "bachelor" is defined as "unmarried man", there is nothing left to explain. Thus, to understand how to answer these meta-explanatory questions, we need to have a grip on the supposed explanations we are investigating.

The main distinction to be made here is between constitutive and non-constitutive explanations, the simplest case here being the case of non-constitutive explanations like governance. Although I have claimed that the dependence relation involved by governance is not causal dependence, there are certain similarities between governance and causal explanation – first and foremost that both of them can be understood along realist lines, according to which such explanations function by tracking worldly relations, such as causal and non-causal dependence relations. Causal laws, as in laws that that characterize general patterns of causal relations, are thus useful meta-explanatory links in the sense that they help understand how a certain (causal) *explanandum* fits into a more general pattern of causal relations. For instance, this is what happens when we try to explain the explanatory claim in (22) by providing a concatenation of causal laws linking the *explanandum* (the window shattering) and *explanans* (Tim throwing a rock at it).

Something similar happens for governance as well: a meta-explanatory link is provided for claims of governance qua explanatory clarifying how a particular instance of governance fits into a more general pattern of non-causal connections. To anticipate some of the contents of the next section, for instance, we can say that this particular dispositional ascription governs this specific L-instance of a regularity because, as a general principle, L-instances of regularities non-causally depend on dispositions. These meta-explanatory links, I will call Governing Hypotheses. The only relevant difference is that whereas we can safely call the meta-explanatory link for (22) causal *laws*, for governance it would perhaps be best to say governance *hypotheses*. For even though we can assume that both causal and non-causal dependence relations exist, the methodology concerning their discovery are wildly different. Causal laws are not discovered by metaphysicians, whereas the positing of governing relations is an explicitly metaphysical issue (Mumford, 2004, p. 4–8): within the context of my methodology for metaphysics, general governing links are provided as hypotheses for entire body of explanations to be evaluated, in this case: let us suppose that L-instances of regularities non-causally depend on dispositional ascriptions, and see how the overall theory functions.

The quest for meta-explanations of constitutive explanations, however, is fairly different. This is mostly because, I take it, there is no out-there grounding relation that might help us clarify what the explanatoriness of the original claim amounts to. (Didn't I just claim that there are non-causal dependence relations? Yes, but they are non-constitutive in the sense that they do not entail identity between the two terms, whereas grounding is standardly understood as a reductive relation. *Constitutive* explanation has no worldly correlate: the only real element of it is the identity itself, which is *per se* non-explanatory.[12]) At any rate, what is constitutive explanation? In short, a constitutive explanation is a clarificatory direction imposed on an otherwise symmetric claim of generalized identity; according to the neo-Quinean

[12]I am assuming that identity, generalized or otherwise, is real in the sense that $\mathbb{S}(=)$: when Superman = Clark Kent, is like when Batman = Bruce Wayne. This is a standard assumption, but nothing of importance should hinge on it.

methodology espoused in this book, it is the (always in principle defeasible) hope of the metaphysician that the theoretically more virtuous body of explanations, so understood, constitutes an increase in perspicuity. This is a crucial difference with non-constitutive explanations such as causal explanations and governance, which can link equally fundamental pieces of reality – there clearly is no expectation that instances of governance layer reality.

In conclusion, if constitutive explanation has no worldly correlate, how can the Connectivist provide a link to answer meta-explanatory questions? There may be more than one answer here. A postulational definition can clarify constitutive explanations – as perhaps is the case for (21) above – but this needn't be always the case. Consider another case of supposedly constitutive explanation for which I am partial, viz. the one of the Ostrich Nominalist (as per Chap. 4):

23) The ball instantiates the property of being red because$_C$ the ball is red

which is supposedly constitutive to the extent that.

24) The ball instantiating the property of being red *just is* the ball being red

Now, what kind of connecting principle is bridging *explanans* and *explanandum* in (23)? Again, it needn't necessarily be a definition: to provide an exhaustive definition of a (potentially obscure) term of art such as "instantiation" seems like too daunting of a task to make (23) work. The point is another, and that is, however we want to define or characterize the concept of instantiation – in reality, the instantiation of the property being red ultimately amounts to the ball being red. In other words, what (23) is claiming, is that what is being expressed in the left-hand side is more perspicuously expressed by the right-hand side; two subsentential components disappear in the *explanans*, "instantiates" and "the property of", thus (23) submits that either one of the two is not joint-carving; instantiation, or propertyhood, is not part of the structure of reality. Other instances of constitutive explanation follow the same pattern.

Let us introduce another term of art for the relevant kind of meta-explanatory link, and see if we put some meat on these bones and characterize it. The term of art is: Constitutive Hypothesis. E.g., the Constitutive Hypothesis clarifying the constitutive explanation in (23) might have the following content; for an arbitrary predicate P:

25) For any x, if x is P, then x instantiates the property of being P

Just like a Governing Hypothesis, a Constitutive Hypothesis is a metaphysician's postulation, to be corroborated by the successes of the theoretical body in which it is embedded. But there are differences as well. Unlike a Governing Hypothesis, it does not postulate the presence of an out-there relation that a successful constitutive explanation is tracking. Neither a Constitutive Hypothesis a definition or clarification of a concept or a predicate, for it is not about concepts or predicates in the first place: the Constitutive Hypothesis whose content is (25) is not about the concept or predicate "instantiates" – it is about instantiation itself (and perhaps propertyhood as well). And although constitutive explanation entails generalized identity, neither can the Constitutive Hypothesis be constituted by claims of identity, e.g., that to be P just is to instantiate the property of being P. As true as that might be, that cannot be the right underpinning for the constitutive explanation in (25) for unlike identity, explanation only runs in one direction. For the Ostrich Nominalist, it is simply false to claim that the ball is red $because_C$ the ball instantiates the property of being red.

Given my refusal to layer reality with grounding, it is difficult to express in clearer terms what a Constitutive Explanation is, as a meta-explanatory link about constitutive explanation. But it needn't be impossible. Remember that on the present account constitutive explanations are a mostly epistemic affair. In these cases, one receives an epistemically successful explanation when one transitions from a metaphysically worst way to a metaphysically better way to express the same portion of reality (allegedly – this is an hypothesis, after all). Thus, we can conceive a Constitutive Hypothesis as a hypothesis of *comparative perspicuity* of *explanans* and *explanandum*. As an example, consider our Ostrich Nominalist again, which will utter:

26) The ball instantiates the property of being red $because_C$ the ball is red,

Which is made explanatory by the following Constitutive Hypothesis:

27) It is a Constitutive Hypothesis that: If something is red, then it instantiates the property of being red

(27) shouldn't be understood as expressing metaphysical grounding, nor causation, nor a conceptual definition, nor identity (generalized or otherwise); (27), as a meta-explanatory link for a Constitutive Explanation simply is a way of expressing the fact that there is a pattern of comparative perspicuity between pairs of sentences such that, e.g.:

28) "The ball is red" is more perspicuous than "The ball instantiates the property of being red"
29) "The car is red" is more perspicuous than "The car instantiates the property of being red"
30) "The table is red" is more perspicuous than "The table instantiates the property of being red"4etc...

And so on for everything within the range of the quantifier in (27). This is the general pattern of comparative perspicuity in which (26) fits, as a genuine constitutive explanation.

Finally, I would like to dispel a possible temptation, that of considering Constitutive Hypotheses as *laws of metaphysics*. As hinted before, one of the options of the Connectivist as in Dasgupta (2014b, p. 568), is to take the meta-explanatory links for explanations to be veritable laws of metaphysics: just as there are there are physical laws backing certain kinds of explanations, there may be metaphysical laws back other kinds of explanations – the same kind of laws, perhaps, that delimitate the boundaries of the much-discussed "metaphysical modality"; as a matter of fact, a small literature has evolved about the nature of these alleged laws, as clearly summarized in Wilsch (2020).

That said, I am skeptical about the notion of a law of metaphysics. Paradigmatically, Schaffer (2017, p. 302) claims that "if there are metaphysical explanations, then there are laws of metaphysics".[13] This may very well be true, but I wonder about the truth of the antecedent. Throughout this book I have argued in favour of non-causal and constitutive explanations of various kinds; such notions have been characterized (to some extent), and specific examples have been discussed which were relevant for the purposes of this book. I have not argued for the far more general claim that there are "metaphysical explanations"; as sceptics about metaphysical grounding have been very clear about (Wilson, 2014), many relations in metaphysics may be explanatory in character; if "metaphysical explanation" is however used in a narrower sense, and not simply as a stand-in for any explanation pursued by the discipline of metaphysics, then something positive ought to be said about it to properly distinguish it along the varieties of explanation. Lacking that, the notion of a "metaphysical explanation" is poorly defined, and the resulting notion of "law of metaphysics" is equally poorly defined. The varieties of explanation explored throughout this book are, I hope, somewhat understood; so to characterize the corresponding principles as "laws of metaphysics" would not be helpful in the slightest.

8.4 Meta-Explanatory Questions about Dispositions

In its humblest form, the current answer to meta-explanatory questions of any stripe consists in providing additional information as to how exactly certain allegedly explanatory claims can in fact be explanatory. According to the Connectivist, seeing how *explanans* and *explanandum* fit into a pre-existing and more general connection is our best bet. But how does this approach measure with respect to the explanations provided throughout this book? What kind of meta-explanatory links should be provided? And where do the explanations stop?

8.4.1 Dispositions and their Explanations

If we are on the right tracks, ordinary dispositional ascriptions govern non-vacuous L-instances of non-accidental regularities. So, an instance of governance, per Chap. 7, could be:

31) S&Pa → Ma because$_G$ $D_{S\&P\text{-}M}$a

By way of necessitation:

[13] Also see Wilsch (2016).

32) $\Box(D_{S\&P\text{-}M}a \rightarrow (S\&Pa \rightarrow Ma))$.

To answer a meta-explanatory question concerning (31) requires a completion for:

33) $(S\&Pa \rightarrow Ma\ \text{because}_G\ D_{S\&P\text{-}M}a)\ \text{because}_C \ldots$

In other words: what does the explanatoriness of $D_{S\&P\text{-}M}a$ with respect to $S\&Pa \rightarrow Ma$ ultimately consist into? And thus, indirectly, why the necessitation in (32)? The Connectivist may offer a Governing Hypothesis whose content is:

34) $\forall x(D_{S\&P-M}x \rightarrow (S\&Px \rightarrow Mx))$

I would like to make something clear: (34) is not itself the Governing Hypothesis, but merely *its content* – much in the same way a usual description such as "the magnitude of the electrostatic attraction between charges is directly proportional to the product of their magnitudes and inversely proportional to the square of the distance between them" is not by itself Coulomb's Law, but merely its content. The distinction is not superfluous. The explaining, at least the kind of explaining non-Humeans are usually thinking about, is not simply achieved by describing the content of a law, but through the statement that *that* is content of a law; viz. not simply by describing a regularity, but by stating that that is not an accidental regularity; for a similar point concerning non-causal and non-scientific explanation, see Wilsch (2020, p. 426). The point generalizes. Our English-struggling Maria not understanding why "Tim is a bachelor because Tim is an unmarried man" will not be satisfied – and will not understand what kind of explanation we are dealing with–, simply by hearing that for every x, if x is a bachelor, then x is a married man (after all, why would that be true?). She will only be satisfied by hearing that *it is by way of a conceptual definition (of the expression "bachelor")* that if x is a bachelor, then x is a married man. Only then the relevant explanation and the correspondent modality become clear.

The same goes here. To do the explaining is the claim that (34) is the content of a Governing Hypothesis, viz:

35) It is a Governing Hypothesis that: $\forall x(D_{S\&P-M}x \rightarrow (S\&Px \rightarrow Mx))$

Which regiments a (supposedly) general pattern of governance to which (31) must conform. In other words: one explains a particular instance of governance by checking whether it conforms to a more general pattern of non-causal dependence relations – much in the same way one explains how a certain event causally explains another by checking whether, and how, their connection conforms to a more general pattern expressed by a causal law.

Let us now consider the following constitutive explanation of laws, as per Chap. 7:

36) L: $\forall x(S\&Px \rightarrow Mx)\ \text{because}_C\ \Box\forall x(Px \rightarrow D_{S\&P-M}x)$

Informally: that it's a law that S&P things are M is constitutively explained by the fact that necessarily, everything that is P, is also $D_{S\&P\text{-}M}$. A meta-explanatory question concerning this explanation is, very simply, an iterated request of constitutive explanation, viz. a completion for:

37) (L: $\forall x(S\&Px \rightarrow Mx)$ because$_C$ $\Box\forall x(Px \rightarrow D_{S\&P-M}x)$) because$_C$ ...

The Connectivist will need a Constitutive Hypothesis to fill in the blanks of (37); but such a Constitutive Hypothesis, given the nature of our analysis of nomicity as a modally robust correlation between dispositions and properties, expressed predicatively, will need to be expressed at the second-order. For example, the content of this hypothesis will be something along the lines of:

38) $\forall S \forall P \forall M\ (\Box\forall x(Px \rightarrow D_{S\&P-M}x) \rightarrow L{:}\ \forall x(S\&Px \rightarrow Mx))$

Again, (38) is not by itself the Constitutive Hypothesis, but merely its content. The Constitutive Hypothesis which has (38) as its content expresses a general pattern of comparative perspicuity between nomic talk and a specific kind of property & disposition talk.

Now, given the pre-emptive clarification on the nature of the explanatory notions involved, meta-explanatory links such as above are much more specific that the vague "it is a metaphysical law that..." envisioned by Dasgupta. Such principles are different, and they function differently, because governance and constitutive explanation are two very different beasts. At any rate, I still think that there is a sense in which they both qualify as instances of what Dasgupta considers a form of "nomic connectivism", if anything because they do not fall in any of the other categories: they are not meant to be conceptually clarificatory, nor they are about essences, nor they are merely modal connections. At the pain of substituting a vague label with an even vaguer label, we could classify (34) and (38) as instances of "Metaphysical Connectivism", with the assumption in mind that such meta-explanatory links in metaphysics do not all serve to function a unified notion of "metaphysical explanation".

8.4.2 *Essentialist Explanations*

A reader familiar with recent development in power metaphysics may feel that something is missing in the explanations provided so far, and namely: the essentialist aspect. For it is a far more common claim for recent friends of dispositions that things behave the way they do, not simply because they are so-and-so disposed, but because they property they have such-and-such a dispositional nature; according to Hildebrand (2020, p. 4) "the behavior of massive objects is not imposed on them from the outside by a god or an external relation like nomic necessitation; rather, the behavior flows from the very nature of mass itself". This recontextualizes meta-explanatory questions in a specific way: perhaps meta-explanatory questions should receive more substantial answers, more informative answers telling us something interesting about the nature of dispositional properties, and how exactly they provide the explanations in the way they do: loosely speaking, for a dispositional essentialist to provide meta-explanatory answers may ultimately amount to dissecting the nature of dispositionality to find some kind of internal mechanism that makes it explanatory in the first place. For a more specific example, consider the meta-explanatory claim concerning the necessitation in Bird's (2007a) second-order essentialism

15) $\Box(SM((S,P),M) \rightarrow \forall x(Sx \& Px \rightarrow Mx))$,

where SM((S,P),M) is the second-order stimulus-manifestation relation constitutive of a potency's essence, is usually understood as explanations concerning the essence of a property: why does the fact that S, P, and M are related by the second-order essential relation SM relation makes it for a regularity? Accordingly, Tugby (2012, p. 725) describes the meta-explanatory difficulty highlighted in Barker and Smart (2012) as follows:

> "we are still entitled to ask what precisely it is about the natures of different kinds of property instantiation which allows them to determine different kinds of causal behaviour. [. . .] The answer to this last question cannot be entirely straightforward for dispositional essentialists of a monistic disposition like Bird. The problem may be put as a dilemma. An obvious explanation as to why instances of different properties give rise to the specific causal behaviours that they do, is that different properties have different inherent or intrinsic natures."

Let me be clear that my objective here is not to solve the difficulty for Bird's dispositional essentialism; or any dispositional essentialism for that matter. I would rather expand on the fact that, given that some kind of essentialism is such a common component in power metaphysics, it may come easy for friends of powers to answer the meta-explanatory questions within my account with essence-based explanation. Essentialism applied to Dispositional Reality, as I conceive it, surely must receive some adjustments; for example, there is no dispositional property qua entity for this essence to be an essence of; so essentialism is better understood through a shift from *objectual* to *generic* essence, as in Vogt (2022). For example, we could deploy a generic essence operator such as "$\boxdot_P$" which reads "it is by virtue of by virtue of what it is to be P", or "it is by virtue of the nature of P"; it does not presuppose that we are expressing the essence of some property-qua-entity. Let us suppose for the sake of the argument that $\boxdot_P$ is perfectly understood, or a primitive of the theory. When applied to dispositional predicates, $\boxdot_P$ is what many friends of dispositions may have in mind when offering essence-based dispositional explanations.

In this section, I will argue against this strategy. As a matter of fact, I don't believe that operators like $\boxdot_P$ are adding anything to the dispositionalist's explanations; as theoretically costly idle wheels, they should be expunged from the account. I am not the first one to claim that "the notion of essence [. . .] is explanatorily inert" (Schaffer, 2017, p. 312), but my observations are peculiar to the account of dispositionality presented so far, and thus are worth discussing.

The simplest way to incorporate our essence operator in a meta-explanatory answer to claims of governance may very well be the following:

39) $\boxdot_{DS\&P\text{-}M}$ ($D_{S\&P\text{-}M}$a → (Sa&Pa → Ma because$_G$ $D_{S\&P\text{-}M}$a))

Which reads, informally: it is in virtue of being so-and-so disposed that, if *a* is so-and-so disposed, then *a* behaves in such-and-such a way because (governance) it is so-and-so disposed. A claim such as (39) is the result of combining my approach to dispositionality with the kind of essentialist variant to the Straightforward Solution proposed in Bennett (2017, p. 196). The deployment of generic essences completely dispenses with the need of reified dispositional properties; furthermore, as for any variant of the Straightforward Solution, this kind of answer to meta-explanatory questions is recursive and may solve any and all explanatory regresses without resort to brutalism.

These are the virtues of the approach. Unfortunately, the main criticism against the Straightforward Solution, as above, remains. The "explanation" provided in (39) is not very informative, if anything because it doesn't say anything about this essential aspect of dispositionality that may warrant the nested governance claim. We are not offered any additional information, beyond the addition of the $\boxdot_{DS\&P\text{-}M}$ sentential operator, that may make (39) genuinely explanatory. In a way, claims like (39) merely announce an essentialist explanation, without actually providing it: for we are told that it is in virtue of the nature of being so-and-so that such-and-such occurs, but we are never told what exactly this nature is, that makes it so. Incidentally, this difficulty generalizes for many similar versions of essence-based dispositionalist explanations; for example, according to Tugby (2012, p. 727) the dispositional essentialist can in fact solve the Barker&Smart argument by resort to what he calls "qualitative" nature of dispositional properties.

> If property instantiations were taken to have a qualitative aspect, they could then be said to have inherent natures which could serve, in principle, to explain the differences in first-order causal behaviour. The qualitative differences among things would, in short, replace the second-order [SM] relations in the ultimate *explanans* for why first-order entities are disposed to bring about the new property instantiations that they are.

Notice the lack of information regarding this supposedly "qualitative aspect" to the nature of dispositional properties. It is proclaimed that such a nature is supposed to explain the disposed object's behaviour, but since no further detail is provided, the explanation never actually occurs. This is not how essences explain: Socrates' human nature may very well explain why Socrates is disposed to think rationally – but only when this human nature is explicitly articulated as (say) being a rational animal, or something along these lines; without this specification, the supposed explanation is but an empty gesture. In this case, Tugby seems to be thinking that since the explanatory burden is lifted from Bird's SM relation, the regress is halted. But the recursive nature of meta-explanatory questions simply underlies a request for further explanatory information whenever it appears to be missing. In this case: *why* is it exactly that Tugby's "qualitative nature" of properties explain lower-level regularities? Unfortunately, no answer is forthcoming.[14]

[14] I suspect that certain friends of dispositions will be sympathetic to the claim that (dispositional) properties have not simply a qualitative nature, but a *primitive* one as well; as such, there is no substantial answer to my question. I will come back to this kind of primitivism in the next section.

In fact, things for (39) are even worst when we consider the possibility of an explanatory regress. For either (39) is explanatorily primitive, thus destroying any hope of receiving any further explanation down the line, or it is not, and the recursive aspect of the Straightforward Solution kicks in, such that the same strategy can be applied to (39) itself. We then obtain:

40) $\Box_{DS\&P\text{-}M}$ ($\Box_{DS\&P\text{-}M}$ ($D_{S\&P\text{-}M}a \rightarrow$ (Sa&Pa $\rightarrow$ Ma because$_G$ $D_{S\&P\text{-}M}a$))).

In a way, this option is even worst: in the recursion, the answer is eternally announced, and never provided.

The intended solution for this difficulty is Connectivism. So the essentialist could try to offer some kind of meta-explanatory linking principle based on essence; for instance, if we are trying to offer completions for the aforementioned.

33) (S&Pa $\rightarrow$ Ma because$_G$ $D_{S\&P\text{-}M}a$) because$_C$…

The Connectivist could provide the following meta-explanatory claim:

41) $\Box_{DS\&P\text{-}M}$ $\forall x(D_{S\&P-M}x \rightarrow (S\&Px \rightarrow Mx))$

That is to say: it is by virtue of being so-and-so disposed that things so-and-so disposed behave in such-and-such a manner. This variant of Connectivism functions by providing essence-based meta-explanatory links, as explored in Dasgupta (2014b, p. 566–567) and Wallner (2021, p. 13–14). Much in the same way, we could provide other essence-based versions of Connectivism, e.g.: it is by virtue of being crimson, that if something is crimson, then something is coloured as well.

This may seem like a step forward: after all, we are saying *something* about the relevant essences. But unfortunately, this isn't enough to make the deployment of essence worth it. After all, the content of (41) is exactly the same as the content of the Governing Hypothesis in (35), the only difference being that such a content is now prefaced by our operator of (generic) essence. Furthermore, I might add, it is not such an addition to make (41) explanatory, since (35) – if the Connectivist is on the right track – is just as explanatory, by providing a relevant link between *explanans* and *explanandum*. If the switch to the essence-based link (41) is to be motivated, one has to argue that (35) is less explanatory than (41), viz. that adding "it is by virtue of what it is to be so-and-so disposed that…" by itself adds explanatory content. Although essence-based explanations are very common amongst friends of disposition, this crucial point is never argued for, and, on my part, I don't know how it could be argued for (although, of course, I do not deny *a priori* that it can). More generally, I feel a sense of unease around the usual dispositionalists claim of things happening "because of *the nature of* such-and-such a power", or, similarly, the world being so-and-so nomically ordered "by virtue of *the nature of* the powers instantiated in the world"; for it isn't clear to me how the essentialist aspect is not simply there to make the claim sound more legitimate. Something similar happens, I take it, for the meta-explanatory Constitutive Hypotheses regulating our explanation of laws through dispositions. E.g., we could formulate the claim:

42) $\boxdot_{DS\&P\text{-}M}(\Box\forall x(Px \rightarrow D_{S\&P-M}x) \rightarrow L: \forall x(S\&Px \rightarrow Mx))$

according to which the necessary correlation between having certain properties and being so-and-so disposed, makes it for a law of nature, by virtue of what it is to be so-and-so disposed (perhaps the nature of the property is relevant as well; variants of the above claims are possible). But the same criticism applies here: why does the addition of the generic essence operator makes it for a more explanatory claim than the Constitutive Hypothesis I submitted earlier on this chapter?

Ultimately, as submitted in the beginning of Chap. 7, I would like to keep the number explanatory primitives of my account of Dispositional Reality fairly low. If, *ceteris paribus*, there is no need to add essence-talk, we shouldn't. For if virtuous constitutive explanations are to be our guide to perspicuity, albeit indirectly, that would make for a theory realist not only about dispositions, but about essences, as well; e.g., it would be the case that $\mathbb{S}(\boxdot)$. But I do not wish to argue that $\mathbb{S}(\boxdot)$. And so, unless some reason to think that its involvement makes for explanatory more privileged claims, I think it would be best to avoid essence-based explanations – as common as they are in the current landscape.

8.4.3 Brute Connectivism: Or, the Business of Dispositions

There's a related problem concerning essence-based forms of Connectivism. This has to do with the possibility of recursion concerning meta-explanatory questions, and the ensuing threat of an explanatory regress.

At the heart of the Barker&Smart criticism to Bird was an explanatory regress: the problem was never that the SM second-order essential relation was a bad *explanans*, but that – by Bird's own light– it could not be the final *explanans*. For the same explanation, or necessitation, or inference if you will, remains to be explained concerning the instance of SM itself. Thus, the regress.

The same problem occurs here, concerning the meta-explanatory links of the Connectivist, essentialist or otherwise. Why is it that things so-and-so disposed behave in such-and-such a way? And why should that be in their nature? Why is it that modally linked properties and dispositions make it for laws of nature? And why should that be in their nature? And so forth. This is, once again, where we see the pitfalls of the essentialist framing of Connectivism.

First, suppose that such meta-explanatory links can themselves be explained. It's unclear to me whether Connectivism has built-in answers to these questions: the pivotal claim of the position is that explanations are achieved by subsuming the *explanandum*, a particularized inference, under a more general pattern; but how can that be achieved, when the *explanandum* itself expresses a general pattern, I am not sure. But of course some answer may be eventually offered, perhaps if we think about this non-systematically. It is not my place here to present or discuss such answers. I would rather point out that, just like before, such meta-meta-explaantory answers are not *per se* essentialist in character, unless we decide to explicitly preface them with our "it is by virtue of..." generic essence operator. This only raises the further question: "why is it by virtue of...?". The threat of regress, mind you, is not specific to the essentialist Connectivist (the very same threat looms over my account as well); but the problem is that the addition of such an important theoretical resource such as our sentential essence operator does nothing to assuage the difficulty – because, if the Connectivist is right, one does not explain by mentioning essences, but by providing links; and so, in a way, by the very nature of explanation, meta-explanatory links supersede essence-based explanations.[15]

The other possibility is, of course, that at some point the explanations stop, and we find some explanatorily primitive claims, employing conceptual resources that we will accept as primitives of our account. Explanations, after all, have to end somewhere. I suspect that many dispositionalists would like to end the explanations sooner rather than later; for example, I suspect that a supporter of Tugby's claim that the qualitative nature of properties explains lower-level regularities, will reply to my question "what is it about this qualitative nature that warrants such regularities?" with the simple answer "nothing, really. That's just the way it is". There is a long-standing association between the claim the (dispositional) properties have a qualitative nature, with the claim that properties have a primitive nature, occasionally bolstered by the association between categoricalism and quidditism about properties. At any rate, I suspect that an essentialist Connectivist will want to claim that, dispositional properties having a primitive nature, there is no more explanation available concerning the meta-explanatory links provided; this results in what Dasgupta (2014b) calls Brute Essentialism.[16]

But if nothing is said about this qualitative nature, and how it makes for lower-level regularities, is there anything really explanatory in claiming that they do? Consider again the case of Socrates as before; viz., the claim that Socrates is able to think rationally by virtue of him being human; but now suppose that, when faced with the question "why?" one refuses to further articulate Socrates' human nature, and instead simply answers "because that's just the way it is. Socrates' human nature is a primitive". It would then seem that we have not provided an explanation as to why Socrates' human provides him with the ability to think rationally – we have simply postulated one such explanation. But explanations cannot be postulated. They must be articulated. Out of the analogy, the supporter of Brute Essentialism provides a Governing Hypothesis whose content is, again:

43) $\boxdot_{DS\&P\text{-}M} \forall x(D_{S\&P-M}x \rightarrow (S\&Px \rightarrow Mx))$.

Viz. it is by virtue of the what it is to be so-and-so disposed, that things being so-and-so disposed behave in such-and-such a way. But if we are to be primitivist about this meta-explanatory link, why not cut the middle-man (essence), and simply claim

[15] This observation is, I think, closer to the point made in Schaffer (2017, p. 312–313).

[16] Relatedly, is the same route available for Bird's dispositional essentialism? Can we claim that the SM second-order relation is an explanatory primitive of the account? I will not say more on the topic, as I do not endorse dispositional essentialism.

that the content of our primitive Governing Hypothesis is that things so-and-so disposed behave in such-and-such a way? What is the point of turning it into an essence-based explanation, if no effort is made (and in fact, it is categorically excluded by the primitivist assumption) to articulate such an explanation?

Cutting the middleman is exactly my stance on the matter. Viz., I believe that we ought to accept the following claims, and their contents, as explanatorily primitives:

- It is a Governing Hypothesis that: $\forall x(D_{S\&P-M}x \rightarrow (S\&Px \rightarrow Mx))$
- It is a Constitutive Hypothesis that: $\forall S \forall P \forall M\ (\Box \forall x(Px \rightarrow D_{S\&P-M}x) \rightarrow L\text{: } \forall x\ (S\&Px \rightarrow Mx))$

(More precisely, we should accept the correspondent meta-explanatory links *schema*, with meta-linguistic placeholders in a predicative position). It is crucial, I believe, to investigate the methodological status of these claims, as unexplained "hypotheses". Of course, explanations must end somewhere; if we take seriously the idea of framing the structure of explanation as an <input, link, output> triple, rather than a simple <input, output> function, then we will see that any explanation presupposes a linking principle; we can of course question the explanatoriness of a link, but only at the expense of adding, another, higher-order link. Thus, either at some point we find a primitive link, or infinite links must be provided. But this is too weak of a claim for us to make: we must also offer some motivation as to why explanations stop *here*, instead of anywhere else. I have argued that the meta-explanatory links of the Brute Essentialist are a bad place to stop; now I must argue that the meta-explanatory links provided in my "hypotheses" are a good place to stop.

The key to answering this question, lies, I believe, in the fact that the aforementioned meta-explanatory links are postulated posits, informing a philosophically-informed notion of dispositionality. Conceptual resources are not added to metaphysical theories without also characterizing its connections with other phenomena, otherwise there would be very little to that addition but a label. This line of reasoning is fascinatingly pursued by Schaffer (2016a) with regard to the "inference problem", generalizing to a vast array of other phenomena; e.g. suppose that we intend to introduce the sentential operator $\Box$, not simply at the syntactic level, but interpreted as expressing metaphysical necessity. This would be impossible without the explicit postulation of connecting principles that inform the behaviour of $\Box$ as an operator of metaphysical necessity. For instance, it would probably be necessary to add:

44) $\Box p \rightarrow p$

But the very same nature of (44) as a postulational posit allowing the interpretation of $\Box$ as metaphysical modality, blocks any further meta-explanatory question. Parodying Lewis' formulation of the inference problem against DTA, Schaffer (2016a, p. 580) writes

> Thus imagine someone saying to the modalist who accepts the T axiom, "I cannot see how it could be absolutely impossible to have Box p but not p. How does your Box govern the truths?" Such a person has simply not understood that the modalist has posited something whose work includes underwriting this very inference via T.

The same happens, I claim, for dispositions. The meta-explanatory links listed above specify the connections between dispositions and regularities, and dispositions and laws. Now, it is true that (most) dispositional predicates, unlike □, predate any metaphysical framework in which they might be embedded. But a theory can enrich certain conceptual resources with new content. This point goes back to something I said in Chap. 4, viz. that the recently described distinction between "metaphysically innocent" and "metaphysically loaded" dispositional talk, amounts to a difference in attitude concerning dispositional ascriptions, viz. the difference between taking them as merely true or metaphysically perspicuous. When we take dispositional ascriptions as metaphysically perspicuous, and thus accepting the existence of Dispositional Reality, we hereby also believe that dispositions may offer various explanations, of various kinds, for other phenomena in the nomic family. That is to say, we are adding new content to dispositional ascriptions, something which is occasionally signalled by the switch to a peculiar kind of metaphysically loaded dispositional terminology, most commonly "powers", as recently as in Williams (2019), but also "potencies" (Bird, 2007a), "potentialities" (Vetter, 2015), and many others. The meta-explanatory links above are the postulations which make these additions philosophically meaningful, and not merely the addition of a label: they specify the kind of explanatory jobs that dispositions are supposed to perform.

8.4.4 The Collapse Argument, Again

As any other form of Brute Connectivism, my position suffers from a version of the Collapse Argument. This is clearly observable in the case of constitutive explanation, which plays the role that grounding was playing in the original argument. My constitutive explanations of nomicity was supposed to corroborate the claim that law-statements, albeit true, are not particularly perspicuous ways to talk about reality and its structure. In particular, they alleged that ~$\mathbb{S}$(L:). And yet, if they are based on meta-explanatory link such as the Constitutive Hypothesis above, and if such links are taken to be primitive, we could argue by way of Purity that all of their subsentential constituents are joint-carving. Unfortunately, the law-making operator appears in the Constitutive Hypothesis; thus, $\mathbb{S}$(L:), *contra* our intention.

As hinted before, there is a temptation amongst the Connectivists who accept Trialism to make a stronger claim concerning their meta-explanatory links: not simply that they are ungrounded, but that they are not even apt at being grounded; in the colourful words of Dasgupta (2014b, p. 592) they "are not part of the building itself but are the scaffolding around which the building is built". The idea is intriguing: the idea is that meta-explanatory links such as the Constitutive Hypothesis are not descriptions of reality, which are apt at being more or less metaphysically perspicuous, but rather something else. This is patently *not* the case for Governing Hypotheses; but the Collapse Argument does not arise in such cases, since no allegedly less fundamental notions appear in the *explanandum* of governance; remember again, that governance does not connect the (allegedly) more perspicuous with the (allegedly) less perspicuous. This subsection is thus entirely revolving around constitutive explanation and Constitutive Hypotheses.

If we were able to effectively demonstrate that explaining meta-explanatory links amounts to something like a category mistake, that would effectively demonstrate that this is, indeed, a good place to stop.

How to develop this line of thought? As the reader will have probably noticed, there's plenty of analogies and metaphors to go around: the meta-explanatory links are the "scaffolding" (Dasgupta, 2014b, p. 592), or the "infrastructure" (Fine, 2005b, p. 325–326; Litland, 2020, p. 142) of our descriptions of reality, rather than part of such descriptions. But how can we express these ideas more pertinently? If the structure of explanation is accepted as the irreducible triple <input, link, output>, we could perhaps understand the difference between fundamental sentences (viz. absolutely perspicuous sentences) and meta-explanatory link, as the difference between inference rules and axioms in a deductive system: they are simply different pieces of the machinery. The problem is that, unlike inference rules, our meta-explanatory links are just sentences: and so, why would it be so categorically wrong to speculate on their fundamentality? (And of course – but this is perhaps where the analogy begins to crack–, in a deductive system with can show, with the help of a deduction theorem, that inference rules are always accompanied by conditional axioms: so the difference between the two may not be so sharp).

To solve this problem, remember our previous difficulty in characterizing the nature of Constitutive Hypotheses. Unlike Governing Hypotheses (and other meta-explanatory links such as causal laws), we cannot take Constitutive Hypothesis as expressing any general pattern concerning an out-there worldly relation: a Constitutive Hypothesis is rather expressing a general pattern of relative perspicuity between sentences. Remember our Constitutive Hypothesis:

- It is a Constitutive Hypothesis that: $\forall S \forall P \forall M\ (\Box \forall x(Px \rightarrow D_{S\&P-M}x) \rightarrow L{:}\ \forall x\ (S\&Px \rightarrow Mx))$

If my analysis is correct, this Constitutive Hypothesis ultimately amounts to the claim that, for each predicates S, P, and M which make for a non-vacuously true instance of the conditional:

45) "$\Box \forall x(Px \rightarrow D_{S\&P-M}x)$" is more perspicuous than "$L{:}\ \forall x(S\&Px \rightarrow Mx)$"

But in sentences like (45), and other components of the pattern of comparative perspicuity collectively expressed by our Constitutive Hypothesis, it is fairly easy to see why considerations of Purity do not arise. The supposedly less fundamental notion, our law-making operator L, in fact appears in a "metaphysically opaque" position (see Chap. 4), and thus, whether (45) is absolutely perspicuous or not (something I won't settle for the time being), we cannot apply $\mathbb{S}$ to L. In other words: nomic talk in (45) is not *used*, it is *mentioned* (as metaphysically less perspicuous).

In conclusion, although my position consists in a version of Brute Connectivism, to the extent that I take Governance and Constitutive Hypotheses – as postulations introducing a metaphysically-informed notion of dispositionality within the theory –

as explanatorily primitive, it does not suffer from the Collapse Argument. Namely: the above Constitutive Hypothesis is not talking about laws, but about the law-making operator L; thus, considerations of Purity cannot force us to conclude that nomicity is fundamental, even if the Constitutive Hypothesis itself is.

8.5 Dissecting Dispositionality

I suspect that some readers, after going through so much of this book, will be somewhat disappointed. The source of this disappointment might be my primitivist attitude towards meta-explanatory questions concerning dispositionality: that is to say, by refusing to elaborate what it is for something to be so-and-so disposed, I have effectively made it impossible to explain why, and how, being so-and-so disposed elicit certain behaviours, or enriches reality with a nomic or modal aspect.

This kind of attitude goes hand-in-hand with the hopes of a "dissection" of dispositionality, so to speak, with the idea that disposition-based explanations such as the ones above are black boxes to opened, so as to reveal some kind of internal mechanism or structure of propertied and related components. A perfect way to express this kind of anxiety is by observing how simple, almost trivial, the meta-explanatory links I provided are, when compared, say, with the complex concatenation of causal laws required to make sense of the link between ordinary causal *explanans* and a causal *explanandum* (consider, by way of example, how complicated the meta-explanatory links required to make sense of "the window shattered because Tim threw a rock at it"). My meta-explanatory links have none of that complexity, and as a result, the answers provided throughout this chapter may have an air of banality to them. To borrow a slogan from Schaffer (2016a), I am simply arguing that dispositions govern because it is their business to govern – and that's the end of the story. How does that make for an interesting position? So perhaps a meta-explanatory (and presumably non-causal) version of the famous *virtus dormitiva* objection may at this point raise its head. Now, I may protest that, unlike the Brute Essentialist, I am *not* in fact merely postulating explanations, instead of providing them. I am rather postulating certain principles which warrant explanations. This is a non-negligible difference: what I am claiming is that if certain general patterns hold, then a certain x explains a certain y; this is not so different from what it is usually done with causal/scientific explanations: for only within certain theoretical context, with certain causal laws in place, certain items are apt to be *explanans* and *explananda*. The difference rather lies in the methodology for discovery of the laws regulating the correspondent explanations: empirically discoverable in one case, and metaphysical postulations in the other.

Yet the objection still stands that these postulations are too simple to be explanatory – that contrary to what I claimed, Governing and Constitutive Hypothesis are not comfortable places to stop in the explanatory regress.

I disagree with this assessment, and I stand by the position espoused in this chapter. The reasons are twofold. Firstly, I contend the idea that all explanations deal in the opening of black boxes, viz. that all explanations are mechanistic explanations. Causal explanations usually are, but mathematical explanations are often not. The same goes for constitutive explanations: that the ball is coloured because the ball is crimson, this is a genuinely successful same-level explanation. All in all, it's not clear to me why there should be any minimal requirement of complexity concerning meta-explanatory links, and to assume the contrary is merely to assume that governance and constitutive explanation are like causal explanation in all regards – which, patently, they are not. And so, it doesn't seem too problematic that my meta-explanations are not effectively dissecting dispositionality, as if dispositionality was a causally operative factor in the production of laws of nature, or wholesale regularities; again, it is patently not.

Secondly, I contend the claim that my (admittedly simple) meta-explanatory links fail to characterize dispositionality in any way, and are thus completely trivial. Some dispositional ascriptions are explanatory primitives in my account, in the sense I do not constitutively explain what it is to be so-and-so disposed through other conceptual resources; therefore, no dissection of dispositionality is indeed possible. What I have done, however, is embed dispositional ascriptions in a very specific explanatory structure, to the extent that if something has such-and-such a property, it will be so-and-so disposed; and that if it is so-and-so disposed then it will behave in such and such a way. Laws emerge in a certain way. Other connections are beyond the scope of this book but constitute future directions of enquiry for supporters of Dispositional Reality. This is non-trivial structural content: more specifically, selecting certain items as primitives as opposed to other is not a trivial claim, although that might emerge only by observing the overall structure of the theory as a body of explanations; e.g. the same piece of reality, something being-so-and-so disposed, explains different things in different ways: it governs behaviour and it grounds laws. This makes the theory non-isomorphic with respect to DTA, or other competitors within the non-Humean camp (or outside of it, for that matter); more precisely, it makes dispositionality a explanatorily unifying posit, whose postulation, unlike in scientific disciplines, cannot be judged directly, but indirectly, along the lines sketched in Chap. 3. This is how a metaphysically loaded notion of dispositionality acquires content, beyond the intuitive content which predates metaphysical theorizing. All in all, it's not any internal mechanism, or some primitively dispositional character (whatever that means) to do the work: but rather, it's the relative position of dispositional ideology in a multi-faceted explanatory hierarchy that gives the notion of Dispositional Reality, as a notion of genuine and worldly potentiality, its distinctive flavour.

8.6 Conclusions

According to non-Humeans, it is not a brute fact that the physical world exhibits the nomic order it does, as opposed to another. Furthermore, within that camp, there are many dispositionalists who believe that we can explain why the world exhibits this specific nomic order by observing the nature of the (dispositional) properties that are instantiated in it. However, I have argued that the deployment of essence-talk in this meta-explanatory endeavour is not particularly useful, and that the dispositionalists ought to avoid it. For the supporter of Dispositional Reality, it is dispositional ascriptions that should figure as *explanans* (of various kinds) for a various range of nomic-related phenomena, from non-accidental regularities to laws of nature. Other potential *explananda*, such as causation and modality in general fall beyond the scope of this book, and will surely constitute a direction for future research.

One may ask: how is it that things being so-and-so disposed enriches reality with a nomic or modal aspect? I prefer a Brute Connectivist stance to answer such meta-explanatory questions, which functions by adding primitive Constitutive and Governing Hypotheses as postulational posits concerning the introduction of a metaphysically informed notion of dispositionality into the theory. There's no reductive definition or constitutive explanation that will clarify, in simpler terms, what it is for something to be so-and-so disposed: in this sense, dispositionality is a primitive of the theory. But it is a non-trivial account. Dispositional ascriptions find themselves embedded in an explanatory structure that is very different from the one espoused by, say, supporters of DTA or dispositional essentialists. In this regard, constitutive explanations – guided by hypothesis of comparative perspicuity between *explanans* and *explanandum* – play a methodologically privileged role. For if our theory will be found to be particularly virtuous, our candidates for absolute perspicuity will be vindicated. That will, in turn constitute a justification for taking fundamental dispositional ascriptions as being absolutely perspicuous.

Chapter 9
The Future of Dispositional Reality

Perhaps more than any other form of non-Humeanism, power metaphysics has the reputation of being somewhat obscure. Indeed, lovers of desertic landscapes have many times and at great length voiced their concern that, while venturing in the jungle, they might encounter some dangerously occult power. In response, friends of powers in the last three decades have tried to show that power metaphysics can indeed be treated as a respectable discipline, worthy of investigation and discussion. There are admirable results here, but it is also true that, on certain occasions, power metaphysics may strike the reader as somewhat cryptic. For me, this is a problem of hidden premises. Think about the so-called "Meinongian problem", perhaps explicitly discussed for the first time in Armstrong (1996, p. 79), and discussed multiple times throughout the literature: the problem appears to be that if powers are directed to their manifestations, and if powers can be unmanifested, sometimes powers can be directed to a non-existent or "Meinongian" entity. Various solutions have been presented to resolve this difficulty. But consider the kind of premises one needs to get the problem off the ground. Not only the power itself must be reified as an entity; but manifestations too – albeit it is not always clear kind of thing a manifestation is. The relation between them must be too: talk of directedness must be interpreted so literally that the possibility of unmanifested dispositions has to be framed as the problematic case of a missing relatum. This is a staggering amount of ontology to be read off the truth of sentences like "the surgarcube is disposed to dissolve when immersed in water", even from a dispositionalist who wants to take the sentence as metaphysically revealing. Furthermore, the credibility of these hidden premises is rarely discussed, which makes the very formulation of the Meinongian problem, in an exhaustive way, and so as to make the premises credible and worthy of discussion, a titanic enterprise: to even show that there is a problem is a difficult task. (Now, it is true that the dispositional predicate in "the sugarcube is disposed to dissolve in water" is logically complex, in a way that at least suggests some kind of internal complexity, but there is very little indication that this grammatical complexity ought to be parsed out in the way the metaphysician wishes; or, at least, I haven't found a single instance in which the discussion of the Meinongian argument is

L. Azzano, *Dispositional Reality*, Synthese Library 482,
https://doi.org/10.1007/978-3-031-52625-1_9

prefaced by considerations to this end.) Power metaphysics is teeming with hidden premises; some of these, as I argued in this book, are mired by fallacious reasoning, such as excessive and unnecessary ontologizing. This has a bad consequence, one that is not proper to power metaphysics, but of all disciplines which require more attentive care: a kind of insularity and self-referentiality that, in my opinion, one would be foolish to ignore.

Within the help of the framework provided in this book, we can focus on the core of realism about dispositions – as an inherently explanatory enterprise. There is nothing occult or metaphysically suspicious in the presence of Dispositional Reality in the structure of reality; at least, nothing more occult than the presence of right and wrong, or possibility and necessity, trees, persons, and cat-dogs. All you require is whichever pre-theoretical grasp you have on dispositional concepts, and the postulation of meta-explanatory links as hypotheses about the relation between dispositions and other items in the nomic family.

There is a distinct non-Humean flavour: an explanatory and modal connection between properties and dispositions sanctions the kind of modal fixity that blocks considerations of Ramseyan humility, and furthermore provides a modally strong correlation between how objects are, and how they are disposed to behave – which is what, on this account, laws of nature ultimately amount to. A denier of necessary connections will find this unpleasant; but non-Humeans friends of dispositions should find themselves right at home.

That said, the contents of this book ought not to be taken as gospel, even by its most sympathetic reader. Even within the boundaries of Dispositional Reality, my account is but one of many. The reader might part ways with me several times in the course of the book; for instance, they might not appreciate my flirt with Ostrich Nominalism in Chap. 4, or the distinction between L-instance and S-instances of regularities, and the ensuing treatment of governance in Chap. 7. This is alright. This book ought to be understood as an opening of logical space, rather than the staunch defence of a specific account. Think of Dispositional Reality as a continent, and my account as a tiny base camp on its shores.

Much of this continent, as a matter of fact, still requires to be explored. The most prominent question that still calls to be answered is: why exactly should we believe that there is Dispositional Reality in the first place? No small matter. Within the context of this book, old issues that were once discussed to settle the question of realism for powers are not as conclusive as we once thought, such as the adequacy of the conditional analysis and the Argument from Science. To score a decisive point in their favour, the friend of dispositions requires a full assessment of the theoretical virtues of their account of Dispositional Reality vis-à-vis its rivals. No small matter indeed!

What the friend of dispositions ought to focus on, are the explanations themselves. Some of them have been attempted in these pages. Others will have to wait. A further difficulty that, so far, has not been resolved, is that these explanations ought not be attempted in isolation, investigating the explanatory roles performed by dispositions in a piecemeal fashion. In the evaluation of an account of Dispositional Reality, the possibility of Dispositional Pluralism needs to be taken into account: viz.

the possibility that more than one kind of dispositional talk is metaphysically perspicuous. *Ceteris paribus*, it would be best if Dispositional Reality was a homogeneous realm. But who is to say what reality looks like?

Consider the following case. We want to expand the contents of this book with an explicit dispositional treatment of (say) possibility and necessity. We do not have to start from scratch. In fact, there are pre-existing accounts that we can borrow from; perhaps the more developed amongst them is Vetter's (2015) potentiality account. And so, the friend of Dispositional Reality with a penchant for potentialities might want to investigate the possibility that potentiality-talk is perspicuous, in the sense that Vetter's POT operator, in the intended interpretation and with its own peculiar grammar, is joint-carving. On this approach, we can straightforwardly examine the metaphysical fallout of choices in logical form; potentialities, so understood, are not things (perhaps *contra* Vetter's original intention); the claim rather is that potentiality-based ideology is best suited to express patterns of worldly similarities.

This is all well and good, but where does that leave us? Suppose, for the sake of the argument, that Vetter's treatment of modality is better than any other rival; and also suppose that the explanations offered in this book are better than any other rival. If these advantages, combined, outweigh the loss of dispositional monism (how exactly are the scales to be balanced, also remains to be seen) that would be proof enough that reality is not homogeneously dispositional, but that at least two distinct patterns of similarities can be found related to dispositional talk. Both dispositions and potentialities exist. How likely is that? Unclear. A potential line of reasoning is that given the connection between nomicity and necessity, it is unlikely that their source ought to be disjointed in this way. But this is not by all means conclusive: if a unique *explanans* could be found to perform all explanatory jobs required of dispositions, the justification for its existence would be that much stronger. In an ideal scenario, the interconnectedness between various dwellers of the nomic family can be explained by their relation with the central concept of dispositionality. After all, dispositions are usually conceived by their supporters as a unifying *explanans*, a veritable "cornerstone" of a new metaphysics (Molnar, 2003, p. 186). But we don't know whether we find ourselves in an ideal scenario. If different dispositional *explanans* independently emerge as preferable, we will need to face the possibility that their individual cases will be that much weaker.

I conclude this brief final chapter with some suggestions for future research that have emerged throughout these pages.

9.1 Dispositions and Nominalism

The intersection between dispositionalism and nominalism about properties is a pivotal point of contention only intermittently discussed in the burgeoning literature on dispositional properties.

Friends and foes of dispositions traditionally considered debates on the metaphysics of properties as orthogonal with regard to the question of realism about dispositionality (Armstrong, 2005, p. 312); this occasionally translated to the notion that the dispositionalist's claims, apparently about properties, could be easily translated into nominalistically kosher formulations (Mellor, 1974, p. 157). This was merely a programmatic claim – at least until now. I have tentatively embraced the combination of Dispositional Reality with Ostrich Nominalism, by taking dispositional predication as explanatorily primitive and explain dispositional property instantiations through such predications. The view is not yet articulated, nor evaluated vis-à-vis other versions of dispositional nominalism, as in Whittle (2009), Vetter (2020), and Vogt (2022). But one can try to fill in the details as they please: there are many ways to be nominalists, and many more ways to believe in properties. What is sure, however, is that many supposed intersections between dispositionalism and the metaphysics of properties were based on hidden premises and excessive ontologizing that, for the supporter of Dispositional Reality, do not hold much water; E.g. as stated above, I do not give much weight to the phenomenon of directedness, and thus the argument in favour of Platonism to be found in Tugby (2013) – which crucially relies on it – does not come off as exceedingly compelling.

9.2 Pushings and Pullings

In my attempts to keep my theory pure, I have tried to keep the conceptual resources employed in my disposition-based explanations to a minimum. Mostly, they consist in first-order logic with dispositional predication. But in the previous chapters, I have committed myself to the reality of something else: governance. In other words, I am committed myself to the kind of non-causal dependence underpinning governance, as an objective non-causal and non-constitutive explanation. Keep in mind that to express it I am using an operationalist approach to explanation, as to minimize the ontological and metaphysical import; but governance still is real, like causal explanation and unlike constitutive explanation. This is not entirely surprising, of course: my account of Dispositional Reality is, after all, a brand of non-Humean metaphysics, so it is not surprising that we will find it populated by all kinds of pushings and pullings; but this doesn't mean that is nothing to clarify.

What is more, the idea that "governance governs causation" also requires clarification. Dispositional ascriptions supposedly govern individual regularity-instances, thus regulating the correspondent causal sequence (including the mathematical structure of quantitative changes). For the non-Humean metaphysician, there is an interplay between governance and causation that still needs addressing.

Incidentally, this might be where we can recuperate some considerations by those who believe that grounding is metaphysical causation: perhaps causal and this kind of non-causal dependence are similar creatures, although it is still unclear to me how. And so, it is not clear to me how its addition weighs on the account, and to what extent it is a detachable component of it. This too, of course, will weigh on the final evaluation of the present account of Dispositional Reality.

9.3 Dispositionalism Without Essentialism

Standardly, non-causal explanations provided by dispositions are essential in nature. It is only more recently that some have tried to explicitly shed away this essentialist aspect, as in Tugby (2021) and Hakkarainen and Keinänen (2021). As the reader might imagine, I am happy to join that crowd. Meta-explanatory questions, as we have seen, is where the lure of essentialism becomes almost irresistible ("dispositions are suited to explain such-and-such a phenomenon because they have such-and-such a nature"); however, I have argued that essences are, most often than not, explanatorily inert.

Someone might object: they don't need to be. For one, the appeal to essence in the derivation in Bird (2007a, p. 46) was admittedly not an empty gesture: a potency's essence was explanatory because it was a *counterfactual* essence, and from the counterfactual we could derive the kind of material conditional standardly used to regiment law-statements. The appeal to counterfactuals to characterize the nature of a potency might have been unfortunate (Jaag, 2014), but the general point stands: what if there was a non-trivial and informative way to characterize the nature of a disposition, that would make it suitable to perform certain explanatory jobs? There is little I can say, a priori, about the impossibility of "dissecting dispositionality", to find, ultimately, a different explanatory primitive underpinning the dispositionalist's explanations. This might very well be. But there is also little I am comfortable adding about this open question at such a level of generality.

9.4 The Identity Theory

An increasingly popular position is the so-called "Identity Theory", according to which the dispositional and the categorical are one and the same. The theory promises many boons in philosophy of science and mind, but there are deep worries concerning its viability, and even its very metaphysical intelligibility. In my account of Dispositional Reality there are no properties to identify, nor aspects or sides thereof. Yet, as suggested in Chap. 4, there may be a way to formulate the Identity Theory within the present framework, viz. by employing a deflated operation of generalized identity. This is a hitherto unexplored possibility that is only available within this account, which potentially dispels many of the metaphysical complications which are proper of the account. But it is not entirely functional, either. The most straightforward difficulty, as per Chap. 6, is that properties are usually associated with more than one disposition. Sphericity cannot simply be a disposition to roll down an inclined surface. Electric charge cannot simply be a disposition to attract and repel. And so forth. A collapse looms, on the assumption that generalized identity is an equivalence. If property P is associated to dispositions D_1 and D_2, it cannot be the case that to be P just is to be D_1, and also that to be P just is to be D_2; for it is not patently the case that to be D_1 just is to be D_2. This difficulty originates from the formal features of identity, and is inherited, through basic assumptions, by generalized identity; it is unclear to me it is to be resolved. More generally, it is unclear to me whether supporters of the Identity Theory will welcome my suggestion. As in many other cases, whether this corrected version of the deflated Identity Theory will deliver its many promises, remains to be seen.

References

Armstrong, D. M. (1968). *A materialist theory of the mind.* Routledge.
Armstrong, D. M. (1969). Dispositions are causes. *Analysis, 30*(1), 23–26.
Armstrong, D. M. (1978). *Universals and scientific realism: A theory of universals*. Cambridge University Press.
Armstrong, D. M. (1983). *What is a law of nature?* Cambridge University Press.
Armstrong, D. M. (1996). *A world of states of affairs*. Cambridge University Press.
Armstrong, D. M. (2005). Four disputes about properties. *Synthese, 144*(3), 309–320.
Audi, P. (2012). A clarification and defense of the notion of grounding. In F. Correia & B. Schnieder (Eds.), *Metaphysical grounding: Understanding the structure of reality* (pp. 101–121). Cambridge University Press.
Austin, C., & de la Rosa, L. (2018). Dispositional properties in evo-devo. In L. de la Rosa & G. Müller (Eds.), *Evolutionary developmental biology*. Springer.
Ayer, A. J. (1956). What is a law of nature? *Revue Internationale de Philosophie, 10*(36), 144–165.
Azzano, L. (2019). The question of realism for powers. *Synthese, 196*(1), 329–354.
Azzano, L. (2020). Dispositionality, categoricity, and where to find them. *Synthese, 199*(1–2), 2949–2976.
Azzano, L. (2022). Dispositional array: Why so scared of possible worlds? *Argumentation, 7*(2), 485–503.
Azzano, L. (2023). Individuating powers: On the regress/circularity individuation arguments against bird's dispositional monism. *Ergo: An Open Access Journal of Philosophy, 10*.
Azzouni, J. (2012). Simple metaphysics and ontological dependence. In F. Correia & B. Schnieder (Eds.), *Metaphysical grounding: Understanding the structure of reality* (pp. 234–253). Cambridge University Press.
Barker, S. (2013). The Emperor's new metaphysics of powers. *Mind, 122*(487), 605–653.
Barker, S., & Smart, B. (2012). The ultimate argument against dispositional monist accounts of Laws. *Analysis, 72*(4), 714–722.
Bauer, W. (2011). An argument for the extrinsic grounding of mass. *Erkenntnis, 74*(1), 81–99.
Beebee, H. (2000). The non-governing conception of laws of nature. *Philosophical and Phenomenological Research, 61*(3), 571–594.
Bennett, K. (2011). By our bootstraps. *Philosophical Perspectives, 25*(1), 27–41.
Bennett, K. (2017). *Making things up*. Oxford University Press.
Bernstein, S. (2016). Grounding is not causation. *Philosophical Perspectives, 30*(1), 21–38.
Bhogal, H. (2020a). Humeanism about laws of nature. *Philosophy Compass, 15*(8), 1–10.

L. Azzano, *Dispositional Reality*, Synthese Library 482,
https://doi.org/10.1007/978-3-031-52625-1

Bhogal, H. (2020b). Nomothetic explanation and Humeanism about laws of nature. In K. Bennett & D. Zimmerman (Eds.), *Oxford studies in metaphysics* (Vol. 12, pp. 164–202). Oxford University Press.

Bhogal, H., & Perry, Z. (2017). What the Humean should say about entanglement. *Noûs, 51*(1), 74–94.

Bird, A. (1998). Dispositions and antidotes. *Philosophical Quarterly, 48*(191), 227–234.

Bird, A. (2005). The ultimate argument against Armstrong's contingent necessitation view of Laws. *Analysis, 65*(2), 147–155.

Bird, A. (2007a). *Nature's metaphysics: Laws and Properties*. Oxford University Press.

Bird, A. (2007b). The regress of pure powers? *The Philosophical Quarterly, 57*(229), 513–534.

Bird, A. (2009). Structural properties revisited. In T. Handfield (Ed.), *Dispositions and causes* (pp. 215–241). Clarendon Press.

Bird, A. (2016). Overpowering: How the powers ontology has overreached itself. *Mind, 125*(498), 341–383.

Black, R. (2000). Against quidditism. *Australasian Journal of Philosophy, 78*(1), 87–104.

Blackburn, S. (1990). Filling in space. *Analysis, 50*(2), 62–65.

Bliss, R. (2014). Viciousness and circles of ground. *Metaphilosophy, 45*(2), 245–256.

Bonevac, D., Dever, J., & Sosa, D. (2011). The counterexample fallacy. *Mind, 120*(480), 1143–1158.

Borges, J. (1962). *Fictions*. Grove press.

Borghini, A. (2009). Dispositions and their intentions. In G. Damschen, R. Schnepf, & K. Stueber (Eds.), *Debating dispositions: Issues in metaphysics, epistemology, and philosophy of mind* (pp. 204–219). De Gruyter.

Borghini, A., & Williams, N. (2008). A dispositional theory of possibility. *Dialectica, 62*(1), 21–41.

Bostock, D. (2012). *Russell's logical atomism*. Oxford University Press.

Broad, C. (1935). Mechanical and teleological causation. *Proceedings of the Aristotelian Society, 14*(Supplement), 83–112.

Bueno, O., & Finn, S. (2018). Quantifier variance dissolved. *Royal Institute of Philosophy Supplement, 82*, 289–307.

Bueno, O., & Shalkowski, S. (2009). Modalism and logical pluralism. *Mind, 118*(470), 295–321.

Bueno, O., & Shalkowski, S. (2013). Logical constants: A modalist approach. *Noûs, 47*(1), 1–24.

Bueno, O., & Shalkowski, S. (2015). Modalism and theoretical virtues: Toward an epistemology of modality. *Philosophical Studies, 172*(3), 671–689.

Busse, R. (2021). Asymmetry cannot solve the circularity/regress problem of property structuralism. *Synthese, 199*(3–4), 10685–10720.

Cameron, R. (2008). Truthmakers and ontological commitment: Or how to deal with complex objects and mathematical ontology without getting into trouble. *Philosophical Studies, 140*(1), 1–18.

Campbell, K. (1976). *Metaphysics: An introduction*. Dickenson.

Carnap, R. (1936). Testability and meaning. *Philosophy of Science, 3*(4), 419–471.

Carroll, J. (1990). The Humean tradition. *Philosophical Review, 99*(2), 185–219.

Carroll, J. W. (1994). *Laws of nature*. Cambridge University Press.

Carroll, J. W. (2008). Nailed to Hume's cross? In J. Hawthorne, T. Sider, & D. Zimmerman (Eds.), *Contemporary debates in metaphysics* (pp. 67–81). Blackwell.

Carroll, J. W. (2021). *Laws of nature*. Retrieved from The Stanford encyclopedia of philosophy (Winter 2020 edition), Edward N. Zalta (ed.): https://plato.stanford.edu/archives/win2020/entries/laws-of-nature/

Cartwright, N. (1983). *How the laws of physics lie*. Oxford University Press.

Cartwright, N. (1989). *Nature's capacities and their measurement*. Oxford University Press.

Cartwright, N. (1999). *The dappled world: A study of the boundaries of science*. Cambridge University Press.

Chakravartty, A. (2007). *A metaphysics for scientific realism: Knowing the unobservable*. Cambridge University Press.

Chisholm, R. M. (1946). The contrary-to-fact conditional. *Mind, 55*(220), 289–307.

Choi, S. (2006). The simple vs. reformed conditional analysis of dispositions. *Synthese, 148*(2), 369–379.

Choi, S. (2008). Dispositional properties and counterfactual conditionals. *Mind, 117*(468), 795–841.

Choi, S. (2009). The conditional analysis of dispositions and the intrinsic dispositions thesis. *Philosophy and Phenomenological Research, 78*(3), 568–590.

Coates, A. (2020). Making sense of powerful qualities. *Synthese, 198*(9), 8347–8363.

Coates, A. (2022). Tropes, unmanifested dispositions and powerful qualities. *Erkenntnis, 87*, 2143–2160.

Contessa, G. (2010). Modal truthmakers and two varieties of actualism. *Synthese, 174*(3), 341–353.

Contessa, G. (2013a). Dispositions and interferences. *Philosophical Studies, 165*(2), 401–419.

Contessa, G. (2013b). Does your metaphysics need structure? *Analysis, 73*(4), 715–721.

Contessa, G. (2014). One's a crowd: Mereological nihilism without ordinary-object eliminativism. *Analytic Philosophy, 55*(2), 199–221.

Contessa, G. (2019). Powerful qualities or pure powers? *Metaphysica, 20*(1), 5–33.

Correia, F. (2006). Generic essence, objectual essence, and modality. *Noûs, 40*(4), 753–767.

Correia, F. (2010). Grounding and truth-functions. *Logique Et Analyse, 53*(211), 251–279.

Correia, F. (2016). On the logic of factual equivalence. *Review of Symbolic Logic, 9*(1), 103–122.

Correia, F., & Skiles, A. (2019). Grounding, essence, and identity. *Philosophy and Phenomenological Research, 98*(3), 642–670.

Cortens, A., & O'Leary-Hawthorne, J. (1995). Towards ontological nihilism. *Philosophical Studies, 79*(2), 143–165.

Cross, T. (2005). What is a disposition? *Synthese, 144*(3), 321–341.

Dasgupta, S. (2014a). On the plurality of grounds. *Philosophers' Imprint, 14*, 1–28.

Dasgupta, S. (2014b). The possibility of physicalism. *Journal of Philosophy, 111*(9–10), 557–592.

Dasgupta, S. (2017). Constitutive explanation. *Philosophical Issues, 27*(1), 74–97.

Davidson, D. (1984). *Inquiries into truth and interpretation*. Oxford University Press.

Davies, P. (2007). *The goldilocks enigma*. Houghton Mifflin Harcourt.

deRosset, L. (2013). Grounding explanations. *Philosophers' Imprint, 13*, 1–35.

deRosset, L. (2014). Possible worlds for modal primitivists. *Journal of Philosophical Logic, 43*(1), 109–131.

Devitt, M. (1996). The metaphysics of nonfactualism. *Philosophical Perspectives, 10*, 159–176.

Dorr, C. (2014). Quantifier variance and the collapse theorems. *The Monist, 9*(4), 503–570.

Dorr, C. (2016). To be F is to be G. *Philosophical Perspectives, 30*(1), 39–134.

Doulas, L. (2021). A puzzle about Moorean metaphysics. *Philosophical Studies, 178*(2), 493–513.

Dresner, E. (2006). Davidson's philosophy of communication. *Communication Theory, 16*(2), 155–172.

Dresner, E. (2011). The principle of charity and intercultural communication. *International Journal of Communication, 5*, 969–982.

Dretske, F. (1977). Laws of nature. *Philosophy of Science, 44*(2), 248–268.

Dummett, M. (1981). *Frege: Philosophy of language* (2nd ed.). Harvard University Press.

Dumsday, T. (2015). Why pan-dispositionalism is incompatible with metaphysical naturalism. *International Journal for Philosophy of Religion, 78*(1), 107–122.

Eagle, A. (2009). Causal structuralism, dispositional actualism, and counterfactual conditionals. In T. Handfield (Ed.), *Dispositions and causes* (pp. 65–99). Oxford University Press.

Eddington, A. S. (1928). *The nature of the physical world*. Dent.

Eklund, M. (2008). The picture of reality as an amorphous lump. In J. Hawthorne, T. Sider, & D. Zimmerman (Eds.), *Contemporary debates in metaphysics* (pp. 382–396). Blackwell.

Eklund, M. (2009). Carnap and ontological pluralism. In D. Manley, D. Chalmers, & R. Wasserman (Eds.), *Metametaphysics: New essays on the foundations of ontology* (pp. 130–156). Oxford University Press.

Ellis, B. (2001). *Scientific essentialism*. Cambridge University Press.

Ellis, B., & Lierse, C. (1994). Dispositional essentialism. *Australasian Journal of Philosophy, 72*(1), 27–45.

Emery, N. (2019). Laws and their instances. *Philosophical Studies, 176*(6), 1535–1561.

Engelhard, K. (2021). Dispositions manifest themselves: An identity theory of properties. *Synthese, 199*(5–6), 13497–13522.

Esfeld, M., & Sachse, C. (2011). *Conservative reductionism*. Routledge.

Fara, M. (2005). Dispositions and habituals. *Noûs, 39*(1), 43–82.

Field, H. (1980). *Science without numbers: A defence of nominalism*. Princeton University Press.

Field, H. (1989). *Realism, mathematics & modality*. Blackwell.

Fine, K. (2001). The question of realism. *Imprint, 1*, 1–30.

Fine, K. (2002). Varieties of necessity. In T. Gendler & J. Hawthorne (Eds.), *Conceivability and possibility* (pp. 253–281). Oxford University Press.

Fine, K. (2005a). Tense and reality. In K. Fine (Ed.), *Modality and tense* (pp. 261–320). Oxford University Press.

Fine, K. (2005b). Necessity and non-existence. In K. Fine (Ed.), *Modality and tense* (pp. 321–354). Oxford University Press.

Fine, K. (2009). The question of ontology. In D. Manley, D. Chalmers, & R. Wasserman (Eds.), *Metametaphysics: New essays on the foundations of ontology* (pp. 157–177). Oxford University Press.

Fine, K. (2012). Guide to ground. In F. Correia & B. Schnieder (Eds.), *Metaphysical grounding: Understanding the structure of reality* (pp. 37–80). Cambridge University Press.

Finocchiaro, P. (2021). High-fidelity metaphysics: Ideological parsimony in theory choice. *Pacific Philosophical Quarterly, 102*(4), 613–632.

Forbes, G. (1992). Melia on modalism. *Philosophical Studies, 68*(1), 57–63.

French, S. (2014). *The structure of the world: Metaphysics and representation*. Oxford University Press.

Friend, T. (2016). Laws are conditionals. *European Journal for Philosophy of Science, 6*(1), 123–144.

Friend, T., & Kimpton-Nye, S. (2023). In T. E. Tahko (Ed.), *Dispositions and powers*. Cambridge University Press.

Giannini, G., & Tugby, M. (2020). Potentiality: Actualism minus naturalism equals platonism. *Philosophical Inquiries, 1*(8), 117–140.

Giannotti, J. (2019). The identity theory of powers revised. *Erkenntnis, 86*(3), 603–621.

Giannotti, J. (2021). Pure powers are not powerful qualities. *European Journal of Analytic Philosophy, 17*(1), 5–29.

Goodman, N. (1947). The problem of counterfactual conditionals. *Journal of Philosophy, 44*(5), 113–128.

Goodman, N. (1954). *Fact, fiction & forecast*. University of London.

Griffin, N. (1985). Russell's multiple relation theory of judgment. *Philosophical Studies, 47*(2), 213–247.

Guay, A., Humphreys, P., & Sartenaer, O. (2021). What price changing laws of nature? *European Journal for Philosophy of Science, 11*(1), 1–19.

Hakkarainen, J., & Keinänen, M. (2021). Away with dispositional essences in trope theory. In L. Jansen & P. Sandstad (Eds.), *Neo-Aristotelian perspectives on formal causation* (pp. 106–123). Routledge.

Halbasch, K. (1971). A critical examination of Russell's view of facts. *Noûs, 5*(4), 395–409.

Hale, B., & Wright, C. (2009). The metaontology of abstraction. In D. Manley, D. Chalmers, & R. Wasserman (Eds.), *Metametaphysics: New essays on the foundations of ontology* (pp. 178–212). Oxford University Press.

Handfield, T. (2008). Humean dispositionalism. *Australasian Journal of Philosophy, 86*(1), 113–126.

Harré, R. (1986). Varieties of realism: A rationale for the natural sciences.

Hawthorne, J. (2001). Causal structuralism. In J. Tomberlin (Ed.), *Metaphysics* (Vol. 35, pp. 361–378). Blackwell.
Heil, J. (2003). *From an ontological point of view*. Oxford University Press.
Heil, J. (2005). Dispositions. *Synthese, 144*(3), 343–356.
Heil, J. (2012). *The universe as we find it*. Oxford University Press.
Heil, J. (2016). Truthmaking and fundamentality. *Synthese, 198*(Suppl 3), 849–860.
Heil, J., & Martin, C. (1999). The ontological turn. *Midwest Studies in Philosophy, 23*(1), 34–60.
Heras-Escribano, M. (2017). Non-factualist dispositionalism. *Philosophia, 45*(2), 607–629.
Hicks, M., & Schaffer, J. (2017). Derivative properties in fundamental laws. *British Journal for the Philosophy of Science, 68*(2), *411–450.*
Hildebrand, T. (2013). Can primitive laws explain? *Philosophers' Imprint, 13*, 1–15.
Hildebrand, T. (2020). Non-Humean theories of natural necessity. *Philosophy Compass, 15*(5), e12662.
Hirsch, E. (2002). Quantifier variance and realism. *Philosophical Issues, 12*(1), 51–73.
Hirsch, E. (2008). Ontological arguments: Interpretive charity and quantifier variance. In J. Hawthorne, T. Sider, & D. Zimmerman (Eds.), *Contemporary debates in metaphysics* (pp. 367–381). Blackwell.
Hirsch, E. (2013). Charity to charity. *Philosophy and Phenomenological Research, 86*(1), 435–442.
Hornsby, J. (2005). Truth without truthmaking entities. In H. Beebee & J. Dodd (Eds.), *Truthmakers: The contemporary debate* (pp. 33–47). Clarendon Press.
Hüttemann, A. (1998). Laws and dispositions. *Philosophy of Science, 65*(1), 121–135.
Ingthorsson, R. (2013). Properties: Qualities, powers, or both? *Dialectica, 67*(1), 55–80.
Ingthorsson, R. (2015). The regress of pure powers revisited. *European Journal of Philosophy, 23*(3), 529–541.
Jaag, S. (2014). Dispositional essentialism and the grounding of natural modality. *Philosophers' Imprint, 14*, 1–21.
Jackson, F., Pargetter, R., & Prior, E. (1982). Three theses about dispositions. *American Philosophical Quarterly, 19*(3), 251–257.
Jacobs, J. (2010). A powers theory of modality: Or, how I learned to stop worrying and reject possible worlds. *Philosophical Studies, 151*(2), 227–248.
Jacobs, J. (2011). Powerful qualities, not pure powers. *The Monist, 94*(1), 81–102.
Jenkins, C. (2011). Is metaphysical dependence Irreflexive? *The Monist, 94*(2), 267–276.
Kelly, A. (2013). Ramseyan humility, scepticism and grasp. *Philosophical Studies, 164*(3), 705–726.
Khlentzos, D. (2021). *Challenges to metaphysical realism.* Retrieved from The Stanford encyclopedia of philosophy (Spring 2021 Edition), Edward N. Zalta (ed.): https://plato.stanford.edu/archives/spr2021/entries/realism-sem-challenge
Kim, J. (2010). *Essays in the metaphysics of mind.* Oxford University Press.
Kimpton-Nye, S. (2021). Reconsidering the dispositional essentialist canon. *Philosophical Studies, 178*(10), 3421–3441.
Kistler, M. (2003). Laws of nature, exceptions and tropes. *Philosophia Scientiae, 7*(2), 189–219.
Kistler, M. (2020). Powers, dispositions and laws of nature. In A.-S. Meincke (Ed.), *Dispositionalism: Perspectives from metaphysics and the philosophy of science* (pp. 171–188). Synthese Library.
Korman, D. (2020). *Ordinary objects*. Retrieved from The Stanford encyclopedia of philosophy (Fall 2020 Edition), Edward N. Zalta (ed.): https://plato.stanford.edu/entries/ordinary-objects/
Koslicki, K. (2012). Varieties of ontological dependence. In F. Correia & B. Schnieder (Eds.), *Metaphysical grounding: Understanding the structure of reality* (pp. 186–213). Cambridge University Press.
Lange, M. (2009). *Laws and Lawmakers: Science, metaphysics, and the Laws of nature*. Oxford University Press.
Lange, M. (2013). Grounding, scientific explanation, and Humean laws. *Philosophical Studies, 164*(1), 255–261.

Lange, M. (2016). *Because without cause: Non-causal explanations in science and mathematics*. Oxford University Press.

Laycock, H. (2006). *Words without objects: Semantics, ontology, and logic for non-singularity*. Oxford University Press.

Leuenberger, S. (2014). Grounding and necessity. *Inquiry: An Interdisciplinary Journal of Philosophy, 57*(2), 151–174.

Lewis, D. K. (1983). New work for a theory of universals. *Australasian Journal of Philosophy, 61*(4), 343–377.

Lewis, D. K. (1984). Putnam's paradox. *Australasian Journal of Philosophy, 62*(3), 221–236.

Lewis, D. K. (1986). *Philosophical papers* (Vol. II). Oxford University Press.

Lewis, D. K. (1991). *Parts of classes*. Blackwell.

Lewis, D. K. (1996). Elusive knowledge. *Australasian Journal of Philosophy, 74*(4), 549–567.

Lewis, D. K. (1997). Finkish dispositions. *Philosophical Quarterly, 47*(187), 143–158.

Lewis, D. K. (2009). Ramseyan humility. In D. Braddon-Mitchell & R. Nola (Eds.), *Conceptual analysis and philosophical naturalism* (pp. 203–222). MIT Press.

Linnebo, Ø. (2014). 'Just is'-statements as generalized identities. *Inquiry: An Interdisciplinary Journal of Philosophy, 57*(4), 466–482.

Lipman, M. (2018). Against fundamentality-based metaphysics. *Noûs, 52*(3), 587–610.

Lipton, P. (1999). All else being equal. *Philosophy, 74*(2), 155–168.

Litland, J. (2015). Grounding, explanation, and the limit of internality. *Philosophical Review, 124*(4), 481–532.

Litland, J. (2017). Grounding grounding. In K. Bennett & D. Zimmerman (Eds.), *Oxford studies in metaphysics 10* (pp. 279–315). Oxford University Press.

Litland, J. (2020). Meta-ground. In M. Raven (Ed.), *The Routledge handbook of metaphysical grounding* (pp. 133–147). Routledge.

Locke, D. (2009). A partial defense of Ramseyan humility. In D. Braddon-Mitchell & R. Nola (Eds.), *Conceptual analysis and philosophical naturalism*. MIT Press.

Loewer, B. (1996). Humean supervenience. *Philosophical Topics, 24*(1), 101–127.

Loewer, B. (2012). Two accounts of laws and time. *Philosophical Studies, 160*(1), 115–137.

Lowe, E. (2005). *The four-category ontology: A metaphysical foundation for natural science*. Clarendon Press.

Lowe, E. (2010). On the individuation of powers. In A. Marmodoro (Ed.), *The metaphysics of powers: Their grounding and their manifestations*. Routledge.

Lowe, E. (2011). How not to think of powers. *The Monist, 94*(1), 19–33.

Mackie, J. (1977). Dispositions, grounds, and causes. *Synthese, 34*(4), 361–369.

Manley, D. (2012). Dispositionality: Beyond the biconditionals. *Australasian Journal of Philosophy, 90*(2), 321–334.

Manley, D., & Wasserman, R. (2008). On linking dispositions and conditionals. *Mind, 117*(465), 59–84.

Manley, D., & Wasserman, R. (2011). Dispositions, conditionals, and counterexamples. *Mind, 120*(480), 1191–1227.

Marmodoro, A. (2017). Aristotelian powers at work: Reciprocity without symmetry in causation. In J. Jacobs (Ed.), *Causal powers* (pp. 57–76). Oxford University Press.

Martin, C. (1993). The need for ontology: Some choices. *Philosophy, 68*(266), 505–522.

Martin, C. (1994). Dispositions and conditionals. *Philosophical Quarterly, 44*(174), 1–8.

Martin, C. (1997). On the need for properties: The road to pythagoreanism and back. *Synthese, 112*(2), 193–231.

Martin, C. (2007). *The mind in nature*. Oxford University Press.

Massin, O. (2009). The metaphysics of forces. *Dialectica, 63*(4), 555–589.

Maudlin, T. (2007). *The metaphysics within physics*. Oxford University Press.

Maurin, A.-S. (2019). Grounding and metaphysical explanation: It's complicated. *Philosophical Studies, 176*(6), 1573–1594.

Maxwell, G. (1978). Rigid designators and mind-brain identity. *Minnesota Studies in the Philosophy of Science, 9*, 365–403.

McKitrick, J. (2003). The bare metaphysical possibility of bare dispositions. *Philosophy and Phenomenological Research, 66*(2), 349–369.

McKitrick, J. (2010). Manifestations as effects. In A. Marmodoro (Ed.), *The metaphysics of powers: Their grounding and their manifestation* (pp. 73–83). Routledge.

McKitrick, J. (2013). How to activate a power. In S. Mumford & M. Tugby (Eds.), *Metaphysics and science* (pp. 123–137). Oxford University Press.

McKitrick, J. (2018). *Dispositional pluralism*. Oxford University Press.

Melia, J. (2005). Truthmaking without truthmakers. In H. Beebee & J. Dodd (Eds.), *Truthmakers: The contemporary debate* (pp. 67–84). Clarendon Press.

Mellor, D. (1974). In defense of dispositions. *Philosophical Review, 83*(2), 157–181.

Miller, E. (2015). Humean scientific explanation. *Philosophical Studies, 172*(5), 1311–1332.

Molnar, G. (2003). *Powers: A study in metaphysics*. Oxford University Press.

Mondadori, F., & Morton, A. (1976). Modal realism: The poisoned pawn. *Philosophical Review, 85*(1), 3–20.

Mulligan, K. (2007). Two dogmas of truthmaking. In J.-M. Monnoyer (Ed.), *Metaphysics and Truthmakers* (pp. 18–51). Ontos Verlag.

Mumford, S. (1998). *Dispositions*. Clarendon Press.

Mumford, S. (1999). Intentionality and the physical: A new theory of disposition ascription. *The Philosophical Quarterly, 49*(195), 215–225.

Mumford, S. (2004). *Laws in nature*. Routledge.

Mumford, S. (2006). The ungrounded argument. *Synthese, 149*(3), 471–489.

Mumford, S., & Anjum, R. (2011). *Getting causes from powers*. Oxford University Press.

Newman, A. (2002). *The correspondence theory of truth: An essay on the metaphysics of predication*. Cambridge University Press.

Oliver, A. (1996). The metaphysics of properties. *Mind, 105*(417), 1–80.

Pagin, P. (2006). The status of charity II: Charity, probability, and simplicity. *International Journal of Philosophical Studies, 14*(3), 361–383.

Peacocke, C. (1978). Necessity and truth theories. *Journal of Philosophical Logic, 7*(1), 473–500.

Peacocke, C. (2002). Principles for possibilia. *Noûs, 36*(3), 486–508.

Pechlivanidi, E., & Psillos, S. (n.d.). *What powers are not*.

Place, U. T. (1996). Intentionality as the mark of the dispositional. *Dialectica, 50*(2), 91–120.

Price, H. (1953). *Thinking and experience*. Hutchinson.

Prior, E. (1985). *Dispositions*. Aberdeen University Press.

Psillos, S. (2006). What do powers do when they are not manifested? *Philosophy and Phenomenological Research, 72*(1), 137–156.

Quine, W. V. (1960). *Word and object*. MIT Press.

Quine, W. V. (1973). *The roots of reference*. Open Court.

Raven, M. (2013). Is ground a strict partial order? *American Philosophical Quarterly, 50*(2), 191–199.

Rayo, A. (2013). *The construction of logical space*. Oxford University Press.

Rayo, A. (2014). Reply to critics. *Inquiry: An Interdisciplinary Journal of Philosophy, 57*(4), 498–534.

Reichenbach, H. (1932). *Atom and cosmos: The world of modern physics*. George Braziller, Inc.

Roberts, J. T. (2008). *The law-governed universe*. Oxford University Press.

Robinson, H. (1982). *Matter and sense: A critique of contemporary materialism*. Cambridge University Press.

Rodriguez-Pereyra, G. (2015). Grounding is not a strict order. *Journal of the American Philosophical Association, 1*(3), 517–534.

Rosen, G. (2010). Metaphysical dependence: Grounding and reduction. In B. Hale & A. Hoffmann (Eds.), *Modality: Metaphysics, logic, and epistemology* (pp. 109–135). Oxford University Press.

Ruben, D.-H. (1990). *Explaining explanation*. Routledge.

Russell, B. (1918). *The philosophy of logical atomism* (1985 Edition). Routledge Classics.
Russell, B. (1927). *The analysis of matter*. Kegan Paul.
Russell, B. (1984). Theory of knowledge: The 1913 manuscript. In Elizabeth Ramsden Eames & Kennet Blackwell (Eds.). Routledge, New York.
Russell, J. (2018). Quality and quantifiers. *Australasian Journal of Philosophy, 96*(3), 562–577.
Ryle, G. (1949). *The concept of mind*. Hutchinson &.
Schaffer, J. (2005). Quiddistic knowledge. *Philosophical Studies, 123*(1–2), 1–32.
Schaffer, J. (2008). Truthmaker commitments. *Philosophical Studies, 141*(1), 7–19.
Schaffer, J. (2009). On what grounds what. In D. Manley, D. Chalmers, & R. Wasserman (Eds.), *Metametaphysics: New essays on the foundations of ontology* (pp. 347–383). Oxford University Press.
Schaffer, J. (2010). Monism: The priority of the whole. *Philosophical Review, 119*(1), 31–76.
Schaffer, J. (2012). Grounding, transitivity, and contrastivity. In F. Correia & B. Schnieder (Eds.), *Metaphysical grounding: Understanding the structure of reality* (pp. 122–138). Cambridge University Press.
Schaffer, J. (2016a). It is the business of laws to govern. *Dialectica, 70*(4), 577–588.
Schaffer, J. (2016b). Grounding in the image of causation. *Philosophical Studies, 173*(1), 49–100.
Schaffer, J. (2017). Laws for metaphysical explanation. *Philosophical Issues, 27*(1), 302–321.
Schnieder, B. (2006). Truth-making without truth-makers. *Synthese, 152*(1), 21–46.
Schrenk, M. (2007). Can capacities rescue us from ceteris paribus laws? In M. Kistler & B. Gnassounou (Eds.), *Dispositions in philosophy and science*. Ashgate.
Schroer, R. (2010). Is there more than one categorical property? *The Philosophical Quarterly, 60*(241), 831–850.
Schroer, R. (2013). Can a single property be both dispositional and categorical? The "partial consideration strategy", partially considered. *Metaphysica, 14*(1), 63–77.
Sellars, W. (1962). Naming and saying. *Philosophy of Science, 29*(1), 7–26.
Shoemaker, S. (1980). Causality and properties. In P. Van Inwagen (Ed.), *Time and cause* (pp. 109–135). D. Reidel.
Shoemaker, S. (1998). Causal and metaphysical necessity. *Pacific Philosophical Quarterly, 79*(1), 59–77.
Sider, T. (2011). *Writing the book of the world*. Oxford University Press.
Sider, T. (2013). Symposium on writing the book of the world. *Analysis, 73*(4), 751–770.
Sider, T. (2014). Outscoping and discourse threat. *Inquiry: An Interdisciplinary Journal of Philosophy, 57*(4), 413–426.
Sider, T. (2020). Ground grounded. *Philosophical Studies, 177*(3), 747–767.
Skiles, A. (2015). Against grounding necessitarianism. *Erkenntnis, 80*(4), 717–751.
Skirry, J. (2003). Three kinds of certainty in Moore's defense of common sense. *History of Philosophy Quarterly, 20*(4), 393–406.
Smith, D. (2016). Quid Quidditism Est? *Erkenntnis, 81*(2), 237–257.
Stebbing, S. (1937). *Philosophy and the physicists*. Methuen & Co.
Strawson, G. (2008). The identity of the categorical and the dispositional. *Analysis, 68*(4), 271–282.
Swinburne, R. (1980). Properties, causation, and projectibility: Reply to Shoemaker. In L. Cohen & M. Hesse (Eds.), *Applications of inductive logic* (pp. 313–320). Oxford University Press.
Swoyer, C. (1982). The nature of natural laws. *Australasian Journal of Philosophy, 60*(3), 203–223.
Taylor, J. H. (2013). In defence of powerful qualities. *Metaphysica, 14*(1), 93–107.
Taylor, J. H. (2018). Powerful qualities and pure powers. *Philosophical Studies, 175*(6), 1423–1440.
Thompson, N. (2016). Grounding and metaphysical explanation. *Proceedings of the Aristotelian Society, 116*(3), 395–402.
Tooley, M. (1977). The nature of laws. *Canadian Journal of Philosophy, 7*(4), 667–698.
Tooley, M. (1987). *Causation: A realist approach*. Oxford University Press.
Trogdon, K. (2013). An introduction to grounding. In M. Hoeltje, B. Schnieder, & A. Steinberg (Eds.), *Varieties of dependence* (pp. 97–122). Philosophia Verlag.

Trogdon, K., & Bliss, R. (2021). *Metaphysical grounding*. Retrieved from The Stanford encyclopedia of philosophy (winter 2021 edition), Edward N. Zalta (ed.): https://plato.stanford.edu/archives/win2021/entries/grounding/

Tugby, M. (2012). Rescuing dispositionalism from the ultimate problem: Reply to Barker and Smart. *Analysis, 72*(4), 723–731.

Tugby, M. (2013). Platonic dispositionalism. *Mind, 122*(486), 451–480.

Tugby, M. (2014). Categoricalism, dispositionalism, and the epistemology of properties. *Synthese, 191*(6), 1–16.

Tugby, M. (2016). Universals, laws, and governance. *Philosophical Studies, 173*(5), 1147–1163.

Tugby, M. (2021). Grounding theories of powers. *Synthese, 198*(12), 11187–11216.

Tugby, M. (2022). Dispositional realism without dispositional essences. *Synthese, 200*(3), 1–27.

van Fraassen, B. (1989). *Laws and symmetry*. Oxford University Press.

Vetter, B. (2011). Recent work: Modality without possible worlds. *Analysis, 71*(4), 742–754.

Vetter, B. (2013). Multi-track dispositions. *Philosophical Quarterly, 63*(251), 330–352.

Vetter, B. (2015). *Potentiality: From dispositions to modality*. Oxford University Press.

Vetter, B. (2020). Explanatory dispositionalism: What anti-Humeans should say. *Synthese, 199*(1–2), 2051–2075.

Vogt, L. (2022). Nominalist dispositional essentialism. *Synthese, 200*(2), *156*.

Wadell, H. (1935). Volume, shape and roundness of quartz particles. *Journal of Geology, 43*, 250–280.

Wallner, M. (2021). The ground of ground, essence, and explanation. *Synthese, 198*(Suppl 6), 1257–1277.

Wang, J. (2018). The epistemological objection to modal primitivism. *Synthese, 198*(Suppl 8), 1887–1898.

Warren, J. (2015). Quantifier variance and the collapse argument. *The Philosophical Quarterly, 65*(259), 241–253.

Whittle, A. (2006). On an argument for humility. *Philosophical Studies, 130*(3), 461–497.

Whittle, A. (2009). Causal nominalism. In T. Handfield (Ed.), *Dispositions and causes* (pp. 242–285). Oxford University Press/Clarendon Press.

Williams, N. E. (2011a). Dispositions and the argument from science. *Australasian Journal of Philosophy, 89*(1), 71–90.

Williams, N. E. (2011b). Putting powers back on multi-track. *Philosophia, 39*(3), 581–595.

Williams, N. E. (2019). *The powers metaphysic*. Oxford University Press.

Williamson, T. (2004). Armchair philosophy, metaphysical modality and counterfactual thinking. *Proceedings of the Aristotelian Society, 105*(1), 1–23.

Williamson, T. (2007). *The philosophy of philosophy*. Wiley-Blackwell.

Wilsch, T. (2016). The deductive-nomological account of metaphysical explanation. *Australasian Journal of Philosophy, 94*(1), 1–23.

Wilsch, T. (2020). Laws in metaphysics. In M. Raven (Ed.), *The Routledge handbook of metaphysical grounding* (pp. 425–436). Routledge.

Wilsch, T. (2021). The governance of laws of nature: Guidance and production. *Philosophical Studies, 178*(3), 909–933.

Wilson, J. (2014). No work for a theory of grounding. *Inquiry: An Interdisciplinary Journal of Philosophy, 57*(5–6), 535–579.

Wilson, A. (2018). Metaphysical causation. *Noûs, 52*(4), 723–751.

Wilson, J. (2021). *Determinables and determinates*. Retrieved from The Stanford encyclopedia of philosophy (Spring 2021 Edition), Edward N. Zalta (ed.): https://plato.stanford.edu/archives/spr2021/entries/determinate-determinables/

Wittgenstein, L. (1922). *Tractatus logico-philosophicus*. Routledge & Kegan Paul.

Woodward, J. (2018). Some varieties of non-causal explanation. In A. Reutlinger & J. Saatsi (Eds.), *Explanation beyond causation: Philosophical perspectives on non-causal explanations*. Oxford University Press.

Yablo, S. (2000). A paradox of existence. In T. Hofweber & A. Everett (Eds.), *Empty names, fiction, and the puzzles of non-existence* (pp. 275–312). CSLI Publications.

Yablo, S. (2001). Go figure: A path through fictionalism. *Midwest Studies in Philosophy, 25*(1), 72–102.

Yablo, S. (2014). *Aboutness*. Princeton University Press.

Yates, D. (2013). The essence of dispositional essentialism. *Philosophy and Phenomenological Research, 87*(1), 93–128.

Yates, D. (2018a). Inverse functionalism and the individuation of powers. *Synthese, 195*(10), 4525–4550.

Yates, D. (2018b). Three arguments for humility. *Philosophical Studies, 175*(2), 461–481.

The manufacturer's authorised representative in the EU is Springer Nature Customer Service Centre GmbH, Europaplatz 3, 69115 Heidelberg, Germany. If you have any concerns regarding our products, please contact ProductSafety@springernature.com

Printed and bound by CPI Group (UK) Ltd, Croydon, CR0 4YY
01/10/2026
02225900-0005